The Non-Darwinian ʳolution

THE NON-DARWINIAN
REVOLUTION

✤ Reinterpreting a Historical Myth

PETER J. BOWLER

The Johns Hopkins University Press

Baltimore and London

© 1988 The Johns Hopkins University Press
All rights reserved
Printed in the United States of America

Originally published in hardcover, 1988
Johns Hopkins Paperbacks edition, 1992

The Johns Hopkins University Press
701 West 40th Street
Baltimore, Maryland 21211-2190
The Johns Hopkins Press Ltd., London

∞ The paper used in this publication meets the minimum requirements of
American National Standard for Information Sciences — Permanence of Paper
for Printed Library Materials, ANSI Z39.48-1984.

Library of Congress Cataloging-in-Publication Data

Bowler, Peter J.
    The non-Darwinian revolution: reinterpreting a historical myth/
Peter J. Bowler.
        p.   cm.
    Bibliography: p.
    Includes index.
    ISBN 0-8018-3678-6 (alk. paper)   ISBN 0-8018-4367-7 (pbk.)
    1. Evolution — History.   I. Title.
QH361.B693 1988   88-9738
575'.009 — dc19   CIP

CONTENTS

FIGURES

PREFACE

Anyone familiar with the work I have done over the last few years will, at least with hindsight, see how it has been leading toward this reinterpretation of the Darwinian Revolution. I have ploughed a somewhat lonely furrow, and studying the history of non-Darwinian evolutionism is still not a fashionable occupation, but there is now a substantial enough body of literature to convince anyone that the parts of Darwin's theory now recognized as important by biologists had comparatively little impact on late nineteenth-century thought. I have found myself writing ever more irascible comments about the Darwin industry in recent times, and, looking back, I am inclined to ask why it took me so long to realize that a new image of nineteenth-century evolutionism was emerging from what had begun as a detailed study of ideas once dismissed as dead ends.

When this book was first outlined to the Johns Hopkins University Press, there were several rather critical referees' responses. One was, "Shouldn't the book be entitled 'The Darwinian Non-Revolution'?" I have some sympathy with this suggestion, but in the end I decided that I wanted the term *Non-Darwinian* in the title, and that the emergence of developmental evolutionism *did* constitute a revolution, although of a far less radical kind than the one portrayed in orthodox accounts. A more serious criticism was, "Everyone in the field now knows that much late nineteenth-century evolutionism was non-Darwinian, so why make such a fuss about it?" My response to this is twofold. First, if everyone does realize this, then the realization seems to be taking a long time to sink in properly, since there seems to be no end to the flood of material based on the assumption that the emergence of natural selection *was* the key influence. Second, even if professional historians of biology accept the point, there are plenty of nonspecialists who need to be alerted to the fact that the development of evolutionism can now be interpreted in a significantly different light.

This brings me to my last point. A complete reinterpretation of Darwinism's effect on Western thought could clearly be written at a very detailed level, but this would have produced a long and expensive book that would have reached a much smaller audience. On this point my own inclinations and those of Henry Tom, at the Johns Hopkins University Press, coincided; we both felt that what was needed was a concise treatment aimed at nonspecialists, with a guide to the relevant, more specialized literature. If I have deliberately overstated my case in some respects, it should be remembered that my chief intention is to stimulate debate.

The Non-Darwinian Revolution

Chapter One

THE MYTH OF THE DARWINIAN REVOLUTION

✤ In 1962 Thomas S. Kuhn's *Structure of Scientific Revolutions* popularized the view that a new scientific theory represents a dramatic shift in the way people view the world. Science is not just the patient accumulation of facts; the facts must be interpreted, and the theoretical framework used to impose meaning on the raw data represents a conceptual pattern, or paradigm, that becomes deeply ingrained in the thought-processes of the scientific community. Paradigms inevitably help to specify how scientists visualize the world in which they live, and in many cases the paradigm has implications in areas such as philosophy, religion, and political thought. When these implications are particularly obvious or important, the paradigm may become part of the world view sustaining the culture of the time. The introduction of a new theory is a traumatic event because it requires the destruction of one paradigm and its replacement with another that rests on conceptual foundations that may be incompatible with those currently taken for granted. In the case of theories with broader implications, a scientific revolution may initiate and symbolize a fundamental change in cultural values.

The two best-known examples of such dramatic changes are the Copernican and the Darwinian revolutions. Kuhn himself devoted another book (1957) to the revolution begun by Copernicus's heliocentric astronomy, which destroyed the traditional assumption that the earth was at the center of the physical universe. The term *Darwinian Revolution* was already in use to denote the equally traumatic developments triggered by the introduction of the theory of evolution. Amid the flurry of activity linked to the centenary of the *Origin of Species* in 1959, Gertrude Himmelfarb's *Darwin and the Darwinian Revolution* used the phrase to sum up the transition to a materialist

1

view of human nature which she supposed had been completed by Darwin's book. The phrase has been used in the title of a more recent book (Ruse 1979), and *Revolution in Science* cited the Darwinian episode as a classic example (Cohen 1985, chap. 19).

The Copernican and the Darwinian revolutions are designated by the name of the scientist who introduced the key stimulus for each development. But the status of Copernicus and Darwin within their respective revolutions is very different. It is widely appreciated that later scientists such as Galileo and Newton had to introduce a new physics in order to turn Copernicus's astronomy into a viable world view. Yet Darwin is often seen as having single-handedly introduced and popularized the essentially materialist view of evolution still accepted by modern biologists. Kuhn pointed out that the truly revolutionary aspect of Darwin's theory was not its evolutionism, but its powerful rejection of the old, teleological view of nature (1962, 171–72). Many surveys of Darwin's cultural impact assume that his particular approach to evolutionism had become well established by the time of his death in 1882. Thus Himmelfarb presents Darwin as a figure who merely summed up the prevailing materialism of his time (1959, chap. 20). Another product of the 1959 centenary, John C. Greene's *Death of Adam*, summarized the effect of Darwin's theory in a chapter entitled "The Triumph of Chance and Change." The view that late-nineteenth-century thought had been dominated by a ruthless "social Darwinism" was expressed in Richard Hofstadter's classic survey ([1944] 1955).

The enthusiasm of modern biologists for Darwin's theory of natural selection helps to explain why the Darwinian Revolution is looked upon as a one-man show. There is far more of Darwin's original theory in modern Darwinism than there was of Copernicus's astronomy in the Newtonian world view that completed the Copernican Revolution. Darwin is a hero of modern science because he introduced both the general idea of evolution and the particular mechanism of change still favored by most biologists. Historians of biology have thus tended to discuss the history of evolutionism as though it were essentially the history of Darwinism. This in turn seems to have convinced the cultural historians that a theory as powerful as natural selection must have had an immediate impact on nineteenth-century thought. But even a fairly conventional survey of Darwinism's later history reveals facts that seem inconsistent with these assumptions. Loren Eiseley's *Darwin's Century* (1958) was one of the few products of the *Origin* centenary that took the trouble to follow the story through to the emergence of modern Darwinism. Eiseley singled out one "missing" factor in Darwin's thinking. This was an adequate theory of heredity, which would only be supplied when Gregor Mendel's laws of inheritance were rediscovered in 1900 and used as the foundation for modern genetics. While seeing genetics as merely completing the picture sketched in by Darwin, Eiseley was nevertheless forced to admit that the theory of natural selection had experienced major difficulties in the period before the fusion with genetics gave rise to the modern "synthetic theory" of evolution.

Within biology there was thus room for major doubt about when the Darwinian Revolution had actually been completed. Ernst Mayr, one of the founders of the synthetic theory, accepted that the revolution did not end until the materialist principles inherent in Darwinism had been accepted. At first he argued that this position was reached in 1883, when August Weismann declared that natural selection was the only viable mechanism of evolution (Mayr 1972a, repr. in Mayr 1976). More recently, Mayr again emphasized Weismann's role in the development of Darwinism, but he conceded that in his own time Weismann was "ridiculed for his speculations" (Mayr 1985, 323). In a more general survey, Mayr noted that the theory of natural selection gained very few converts in the late nineteenth century (1982, 514), a view shared by Michael Ruse (1979, 228–33). It begins to look as though the Darwinian Revolution was not such a one-man show after all. Darwin may have hit upon the explanation of evolution favored by modern biologists, but he obviously encountered major difficulties in presenting it to his contemporaries. By any sensible meaning of the term, the *Darwinian Revolution* in biology was not completed until the synthesis with genetics in the 1920s and 1930s.

There is surely a paradox here. Historians concentrating on Darwin's cultural impact seem content with the idea that the materialist view of nature was generally accepted in the late nineteenth century. Yet those who study the development of biology agree that the theory of natural selection — surely the heart of Darwin's materialism — had little effect until the twentieth century. I believe that the potential incompatibility between these two positions has been concealed by a failure to explore the alternatives to Darwinism that flourished within nineteenth-century evolutionism. Historians of biology who admit that the selection theory had little immediate impact have nevertheless continued to focus on the limitations of Darwin's thought, especially his lack of a modern theory of heredity. Because they have dismissed anti-Darwinian evolutionism as irrelevant, they have produced nothing that might persuade the cultural historians to reconsider their assumption that Darwinism dominated the Victorian world view. I suggest that it is unreasonable to believe that a theory that failed to impress the scientists of the time could have brought about a major cultural revolution. Once this point is accepted, one is led to suspect that the traditional interpretation of the Darwinian Revolution is a myth based on a distorted image of Darwin's effect both on science and on the emergence of modern thought.

My conviction that Darwin's role needs to be reassessed has grown out of a decade or more of historical research on the one topic ignored by the conventional interpretation, namely, the non-Darwinian evolutionary ideas now repudiated by most modern biologists. In my book, *The Eclipse of Darwinism*, published in 1983, I described how these ideas flourished within scientific biology in the decades around 1900. The apparently provocative title was based on a phrase coined not by an anti-Darwinian but by Julian Huxley, another founder of the Modern Synthesis. A small but significant number of historical studies are now available to throw additional light on this episode, including some that reveal the very limited appreciation of the selection theory shown by some self-proclaimed Darwinians. More recently, I turned to the strangely neglected area of human evolution (Bowler 1986) only to find that here, too, non-Darwinian ideas had predominated into the 1930s. This work complemented existing studies of the history of anthropology by J. W. Burrow (1966) and George W. Stocking (1968), which had already shown how the idea of cultural evolution emerged independently of Darwin. Robert Bannister's (1979) revaluation of the role played by social Darwinism also suggested that Darwin's influence on social thought might have been exaggerated. The purpose of the present book is to use these and many other studies as the basis for a reinterpretation of nineteenth-century evolutionism which will allow full recognition of the role played by non-Darwinian sources.

Let me first make it clear what I do and do not mean when I assert that the conventional image of the Darwinian Revolution is a myth. I do *not* mean that Darwin's theory constitutes a blind alley along which scientists have been led by their cultural or philosophical preconceptions. On the contrary, it seems clear that Darwin pioneered certain ideas that have proved immensely stimulating to modern biologists. Nor do I want to claim that the *Origin of Species* had no effect at all on nineteenth-century thought. I suspect that the book's influence has been somewhat exaggerated, but I am far more interested in exploring the possibility that the impact of Darwin's theory has been misunderstood. The *Origin* certainly played a role in converting the English- and, to a lesser extent, the German-speaking world to evolutionism. But there were other forces promoting an evolutionary view of nature, often along lines very different from those proposed by Darwin. My suggestion is that Darwin's theory should be seen not as the central theme in nineteenth-century evolutionism but as a catalyst that helped to bring about the transition to an evolutionary viewpoint within an essentially non-Darwinian conceptual framework. This was the "Non-Darwinian Revolution"; it was a revolution because it required the rejection of certain key aspects of creationism, but it was non-Darwinian because it succeeded in preserving and modernizing the old teleological view of things.

Interpreted in this way, the story of nineteenth-century evolutionism centers not on Darwinism (as it is recognized today) but on the emergence of what might be called the "developmental" model of evolution. By stressing the orderly, goal-directed, and usually progressive character of evolution, often through a comparison with individual growth, this model preserves certain aspects of the traditional world view. Moves in this direction were already being made before Darwin was published, and I believe that the developmental model, in one form or another, continued to dominate late nineteenth-century thought. The materialist aspects of Darwin's theory which appeal to modern biologists were not typical of his own time; indeed, they were so radical that hardly anyone could accept them. Darwin achieved the status of a cultural symbol because his book catalyzed the transition to a full-fledged evolutionism within the developmental tradition. It forced the exponents of this tradition to confront the need to reject certain essential aspects of creationism, especially the traditional separation between human beings and the rest of nature. The conventional image of a heated debate centered on the *Origin* reflects the discomfort felt by many conservatives. Recent studies have suggested, however, that the problems were resolved fairly quickly, often by assuming that evolution itself was a purposeful process. The basic transition to evolutionism was thus a rather limited revolution. The antiteleological aspects of Darwin's thinking prized by modern biologists were evaded or subverted by the majority of his contemporaries.

Histories of evolutionism written around 1900 treated Darwin as an important stimulus but tended to dismiss his theory of natural selection as peripheral to the subject's real development. To suggest a return to this perspective today, deliberately excluding the benefits of hindsight, might seem perverse, and yet it would reveal many of the factors that still shape the nonscientist's image of evolution. Only after the emergence of modern Darwinism via the synthesis with genetics did it become clear that Darwin's original theory could be far more useful in biology than his contemporaries had realized. He had been, in conventional parlance, ahead of his time. Mid-twentieth-century historians thus began to treat the *Origin* as a watershed, not only in popular support for evolutionism but also in the development of evolution theory. Taking note of the late-nineteenth-century view of Darwin's achievement allows us to see that the revolution centered on the emergence of Mendelian genetics represents a far more basic transformation of biologists' attitudes. Mendelism did what Darwin could not do: it undermined the plausibility of the analogy between evolution and growth. Shattered by this and other transformations of Western thought in the early twentieth century, the faith in progress that had sustained Victorian evolutionism began to disintegrate. If the memory of Darwin's theory played a role in this later transition, it was but one among many factors. Thus the "Darwinian Revolution" may have been completed by forces that were only indirectly set in motion by the *Origin of Species*.

The details of these proposals will be described in the chapters that follow, but there are two preliminary tasks that must be completed before the reassessment can begin. First I must clarify what I mean by the Darwinian and non-Darwinian approaches to evolutionism. Then I must amplify the hints I have already dropped to build up a convincing explanation of why the non-Darwinian approach has been so systematically ignored by historians seeking to focus attention on Darwin's role.

Darwinian and Non-Darwinian Modes of Evolution

The term *Darwinism* has often been used as little more than a synonym for *evolutionism*. This usage symbolizes Darwin's position as a founding father of the evolutionary approach to nature, but it has generated immense confusion among those who seek to understand how his theory contributed to the growth of evolutionary thought. In its nineteenth-century context, evolutionism, and hence Darwinism, often included a commitment to the inevitability of progress, a position that would be repudiated by modern biologists as a violation of the principles now thought to underlie Darwin's approach to the issue. It must be admitted that the modern interpretation extracts these principles from the corpus of Darwin's work at the cost of ignoring the context within which he strove to popularize his ideas. There are some passages in Darwin's writings that present a very "modern" attitude toward evolution, and some that equally clearly reflect Victorian attitudes and values. But even in Darwin's own time, there were some biologists who defined Darwinism according to the approach suggested in the more radical passages, and it is this definition that survives today. To avoid the need for constant qualification, a survey of the development of evolutionism from the nineteenth into the twentieth century must adopt a consistent definition of Darwinism. Since a theory based on certain aspects of Darwin's thought survives today, the use by historians of a more general (and sometimes contradictory) definition based on the Victorians' understanding of evolution can only invite confusion. Thus when I say that Darwinism was not widely accepted in the late nineteenth century, I mean that those aspects of Darwin's thought most attractive to modern biologists were not accepted at that time.

How then does one define the essence of the Darwinian approach and identify the alternatives that might exist within the broader framework of evolutionism? The mechanism of natural selection is certainly the key to Darwin's most radical insight, but it is essential to define the major differences between this mechanism and alternatives designed to retain an element of teleology and progressionism. The analysis outlined below closely parallels that advanced by Richard Lewontin (1983), although it begins at a different point, and I prefer the term *developmental* for what Lewontin calls "transformational," or non-Darwinian, evolutionism. Significantly, Lewontin sees the developmental approach as characteristic of the pre-Darwinian era. My argument is that we must take account of its continued popularity long after Darwin introduced his own "variational" model.

The most radical aspect of Darwin's approach was his reliance on adaptation as the sole driving agent of evolution. Species change because they must adapt to new environments or because they become more specialized for their existing life styles. There is no other force predetermining their course of development as they respond to the adaptive challenges thrown up by an ever-changing environment. From this assumption follows a whole complex of beliefs about how living things have developed, and it is this complex that represents the true essence of Darwinism. Once adaptation is accepted as the sole directing agent, evolution has to be seen as an irregularly branching tree — not as the ascent of a ladder toward some predetermined goal (Gould 1977a, chap. 6). The uncertainties of migration introduce an element of chance and irregularity into the process. If samples from a single original species migrate to several new locations, each will adapt to its new environment in its own way, and eventually a group of distinct but related species will be formed. Each of the descendant species will then undergo its own evolutionary process, depending on the opportunities for further migration and adaptation that open up to it. Evolution must thus be regarded as a pattern of haphazard branching, with the branches constantly diverging further apart and redividing where possible (see fig. 1). No one branch can

Figure 1. Part of the diagram from Darwin's *Origin of Species*, which illustrates the irregularly branching character of adaptive evolution.

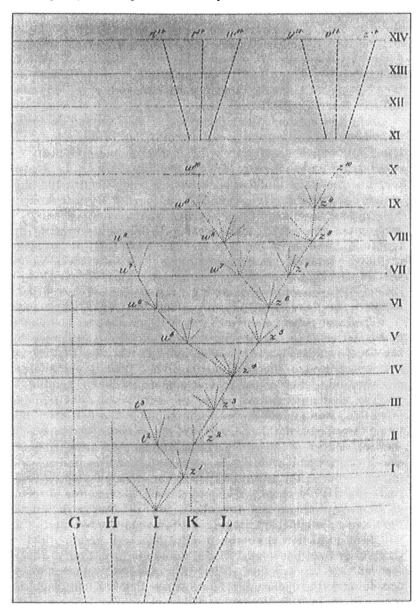

represent the goal toward which all the others should be moving, nor can the stages in the evolution of one branch be used as a hierarchy against which the progress of others can be measured.

Darwin devised the theory of natural selection as an explanation of how populations adapt to changes in their environment. It is not the only conceivable mechanism of adaptation (see below), but it had certain characteristics that drove home the essentially open-ended nature of the Darwinian approach. Natural selection works because, in the struggle for existence, those individuals best adapted to new conditions will survive and reproduce, while others less well adapted will be eliminated. Over many generations, the adaptive character is thus enhanced until eventually the population becomes a new species, incapable of interbreeding with the parent form (if that still exists). The variations from which the environment selects are essentially random and undirected; they are the trivial everyday differences that are used to identify animals and human beings as individuals. The fact that variations are not directed along fixed lines emphasizes the haphazard or open-ended nature of a process governed only by adaptation. This model makes it possible for two originally identical populations exposed to the same environment (for instance, on two very similar islands) to adapt in different ways, because different kinds of favorable variations just happened to appear first in each group. The course of evolution is thus totally unpredictable.

Because the theory of natural selection is now recognized as Darwin's most important discovery, biologists are perhaps inclined to take for granted the wider Darwinian picture of branching, adaptive evolution. If we simply assume that evolution *must* be largely adaptive, then natural selection becomes the essence of Darwinism, and the only conceivable form of non-Darwinian evolution would be an alternative mechanism of adaptation. There is one such alternative, considered from time to time by biologists but now largely rejected as incompatible with modern genetics. This is J. B. Lamarck's theory of the inheritance of acquired characters, proposed in works such as his *Zoological Philosophy* of 1809. Lamarck ignored the possibility of random variation, assuming that in a new environment *all* the individuals will strive to acquire useful new habits. Their bodily structure will be changed by the new habits, just as a weight lifter's arms are developed in response to his unusual pattern of exercise. If such acquired characters are inherited, even to a slight extent, the next generation will be born with a structure that is already slightly modified and will, of course, continue the modification process. Over many generations the species will thus become significantly modified. Lamarck's best-known example (trans. 1914, 122) involves the giraffe.

According to Lamarck, the ancestors of the modern giraffe stretched their necks to reach the leaves of trees; the effect of the stretching, inherited over many generations, accumulated to produce the long neck that now distinguishes the species.

Natural selection and Lamarckism seem to be the only mechanisms of adaptive evolution that have ever been proposed (although, as I shall discuss, there are many others that turn evolution into a nonadaptive process). Lamarckism is the best-known alternative to Darwinism because it has been supported from time to time by writers looking for something more humane than the survival of the fittest. From the time of Samuel Butler (1879), the view that Lamarckism would allow the individual animal to determine its own fate and thus direct the evolution of its species has been urged as an antidote to the materialism of natural selection. Arthur Koestler (1967) was a recent advocate of this position. Yet, as Koestler admitted, the Lamarckian program is not quite so clearly distinct from the Darwinian as Butler once supposed. Selectionists, too, can accept that changes in individual behavior are the starting point for new adaptive trends and that natural selection merely adapts bodily structure to the chosen habits. This is the so-called Baldwin effect, proposed in the 1890s by the psychologist James Mark Baldwin and others (see Baldwin 1902). Within this model, it would be possible for populations exposed to identical conditions on separate islands to choose different habits as a means of adapting to the new environment and thus evolve in different directions. Thus, as Butler himself seems to have recognized, simple Lamarckism does not challenge the belief that evolution is a haphazard and open-ended process.

Lamarckism *did* become part of a more generally anti-Darwinian view of evolution, but this was because it could also be associated with a system that challenged Darwin's reliance on adaptation as the only directing agent. Such an alternative supposes that the course of evolution is predetermined — that a species must evolve through a preordained hierarchy of developmental stages whatever the environments to which it is exposed. There was a strong sense that progress was a force independent of adaptation in Lamarck's original theory (Hodge 1971; Burkhardt 1977), which is why any claim that he and not Darwin was the true founder of modern evolutionism should be viewed with suspicion. In the early nineteenth century, there were many naturalists who drew attention to the alleged parallel between the history of life on the earth (as revealed by the fossil record) and the growth of the human embryo (Bowler 1976a; Gould 1977b). The advance from invertebrates to fish, reptiles, mammals, and, finally, human beings was thought to be mirrored in the purposeful growth of the embryo toward maturity. Small wonder that such an approach saw the history of life on earth as the unfolding of a meaningful plan for reaching creation's predetermined goal. Although at first promoted within a creationist framework by naturalists such as Louis Agassiz, the developmental model seemed ready-made for an evolutionary interpretation in which the advance of life on the earth, like the growth of the embryo, could be seen as occurring through a continuous transformation to higher levels of organization.

This was the view expressed in the first attempt to popularize the idea of evolution in the English-speaking world, Robert Chambers's anonymously-published *Vestiges of the Natural History of Creation* (1844). Although often treated as a simple precursor of Darwinism (Millhauser 1959), Chambers's work advanced a completely non-Darwinian view of organic development (Hodge 1972). For Chambers, the history of life represented the ascent of a preordained hierarchy that could be regarded as a divine plan somehow programmed into nature. Just as the individual organism grows toward maturity, so evolution advances toward its preordained goal. Each step in the ascent represents the addition of a new stage to the growth-pattern of the organism. Variation is not a random disturbance of growth, but the addition of stages to extend the pattern already laid down. Because variation is an addition to growth, the development of the modern embryo will follow the sequence of adult forms through which the species has passed in the course of its evolutionary history. The belief that ontogeny (individual growth) recapitulates phylogeny (the history of the type) thus owes little to Darwinism and is more characteristic of the non-Darwinian, or developmental, view of evolution.

Darwin saw evolution as a branching, open-ended process governed by adaptation and the hazards of migration, but Chambers postulated an inevitable ascent through a fixed hierarchy of organization. Later editions of *Vestiges* invoked parallel evolution along the scale to explain the resemblances that Darwin attributed to common descent (see fig. 2). In Darwin's theory, when a parent species gives rise to several divergent forms, the descendants are still closely related because they share many characters inherited unchanged from the parent. In Chambers's system, similarities between species need not be due to common descent, since many lines of evolution are constrained to advance in parallel through the same hierarchy of developmental stages. Two or more forms will appear to be closely "related" because they have independently reached the same stage of development and have acquired the characters typical of this stage. The

Figure 2. Darwinian and non-Darwinian modes of evolution.

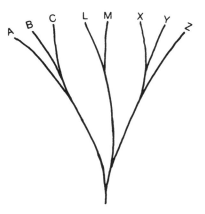

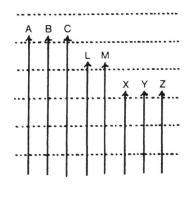

Darwinian Model

Evolution is branching and irregular.

Variation occurs by random mutation that switches individual development into new channels.

Similarity of related forms is due to descent from a common ancestor.

Non-Darwinian Model

Evolution is regular, hierarchical, and preordained.

Variation occurs by the addition of stages to individual development.

Similarity of "related" forms is due to separate lines independently reaching the same stage of development.

hierarchy of typical forms determines the pattern of evolution, and the actual cause of transmutation is of little interest since the end result is preordained by the pattern of development. Chambers hinted that the environment might somehow set off transmutation, but had no real interest in adaptation.

Chambers's theory seems almost to rest on that ancient concept, the chain of being, in which all living forms are arranged into a single linear hierarchy (Lovejoy 1936). It is often supposed that the chain of being was destroyed by early-nineteenth-century naturalists such as Georges Cuvier, who created the open-ended view of natural relationships within which modern evolutionism would eventually flourish (Foucault 1970). Yet the idea of a linear pattern of progress as the central theme (if not the whole story) of the history of life was retained by many post-Darwinian biologists. Side branches were dismissed either as trivial variations on a theme or as the simple preservation of earlier forms into the present. Even when substantial branching was admitted, the pattern of development within each branch was often assumed to follow a predetermined line toward a predictable goal. Instead of dismissing Chambers as a somewhat confused precursor of Darwin, we should recognize him as an early exponent of the developmental viewpoint that was to dominate most late-nine-teenth-century thinking on evolution.

Many of the biologists who proclaimed themselves followers of Darwin remained faithful to some aspect of the developmental view. Ernst Haeckel's pioneering efforts to reconstruct the history of life on earth were based on a progressionist vision of evolution with a main line of development aimed at mankind. It was Haeckel, not Darwin, who popularized the recapitulation theory as an integral part of late-nineteenth-century evolutionism. Although Haeckel claimed that growth merely followed the path established by evolution, his use of the analogy with growth seemed to carry the implication that evolution, too, was a purposeful process. He appealed openly to La-marckism, not because he was interested in adaptation, but because acquired characters are additions to growth, and their inheritance would create a framework of plausibility for the recapitulation theory. Although not an exponent of parallel evolution, Haeckel treated many lower forms as "living fossils" — scarcely changed relics of earlier stages in the main line of human evolution. The combination of Lamarckism, the recapitulation theory, and the idea of linear evo-lution can also be seen in the thought of the group of biologists known as the American neo-Lamarckians. But Lamarckians such as Edward Drinker Cope *did* invoke parallel evolution, and they insisted that the pattern of a group's evolution was not determined solely by ad-aptation. Their explicitly anti-Darwinian version of evolutionism be-came increasingly popular in the last decades of the nineteenth cen-tury.

The term *evolution* originally referred to the growth of the em-bryo and was seldom used by Darwin to denote the transmutation of species (Bowler 1975). We now routinely speak of the theory of ev-olution because the idea was popularized largely via the develop-mental model. In fact, the term was first applied in a broader context by social thinkers, many of whom adopted a developmental view of cultural evolution. Anthropologists such as E. B. Tylor and L. H. Morgan believed that all races advance through the same pattern of cultural stages, invoking a parallelism exactly equivalent to that in Chambers's biology. Modern "primitives" belonged to retarded lines of cultural development which remained at a stage through which the white race had passed at an early period in its history. The so-ciologist Herbert Spencer adopted a more flexible view of cultural evolution, yet still treated free-enterprise capitalism as the goal to-ward which all social progress was aimed. The Victorians welcomed Spencer's emphasis on progressive evolution through the struggle for existence, but they did not share Darwin's view that the purpose of struggle was to eliminate those who are congenitally maladapted to their environment. The elimination of the unfit was an unfortunate byproduct of a process whose real purpose was to stimulate all individuals and all races to maximize their self-development. Such ideas may certainly be found in Darwin's own writings — but not in those passages to which biologists now turn in their search for the origins of modern "Darwinism."

The Darwin Industry and Its Opponents

Although some work has been done on the history of non-Darwinian evolutionism, it pales to insignificance beside the efforts devoted to Darwin and his theory. The concentration of scholarly activity on this topic has become so intense that the "Darwin industry" has become a familiar term. Over the last few decades, Darwin's notebooks have been edited and published (Darwin 1959a, 1960–61, 1963, 1967, 1974, 1979, 1987). There is a variorum text of the *Origin of Species* (Darwin 1959b), and the first edition has been supplied with that indispensable tool of modern scholarship, a concordance (Barrett, Weinshank, and Gottleber, 1981). Darwin's lesser-known books on subjects such as earthworms have been reprinted with introductions by eminent modern biologists (Darwin 1985b). The project to publish all of Darwin's correspondence has now reached its second volume (Darwin 1985a, 1986). Up to twenty volumes will be needed to finish the job, spread over the rest of this century. So many conflicting accounts of how Darwin discovered his theory have now appeared that a lengthy survey has been published to help the uninitiated find their way into the maze (Oldroyd 1984). There is a new volume of articles on Darwin that runs to 1,019 pages of text (Kohn 1985). The centenary of Darwin's death in 1982 was marked by a flood of academic conferences all over the world (Wassersug and Rose 1984). A lavishly produced commemorative volume was sold at a price of $585 (Chapman and Duval 1982). Darwin has become a big business, as indicated by a flood of coffee-table books (e.g., Moorehead 1969), TV series (e.g., Ralling 1978), and popular novels and biographies (e.g., Brent 1981; Stone 1981; Clark, *Survival,* 1984).

The enthusiasm of today's scholars contrasts significantly with the attitude of an earlier generation of scientists and historians. Of course Darwin was recognized as an important figure by his contemporaries, since no one doubted that the *Origin* had sparked the debate that led to widespread acceptance of evolutionism. Darwin was buried in Westminster Abbey as a symbol of the position he had been accorded by his fellow scientists (Moore 1982a). Assessing the impact of the *Origin* in Darwin's *Life and Letters*, T. H. Huxley drew a ritual comparison with Newton and Faraday; Darwin had extended science into an entirely new area (Darwin 1887, 2:179). A substantial volume was published to mark the *Origin's* fiftieth anniversary (Seward 1909), by which time the belief that Darwin had precipitated a revolution in biology was already commonplace (Judd 1911). Yet attitudes toward the theory of natural selection were somewhat more ambivalent. Many biologists thought that Darwin had been mistaken in pinning his faith on this mechanism. Books such as G. J. Romanes's *Darwin and after Darwin* (1892–97) and Vernon Kellogg's *Darwinism Today* (1907) were concerned mainly with the difficulties facing the selection theory. There was even a book entitled *At the Deathbed of Darwinism* (Dennert 1904). As late as 1929 the respected historian of biology Erik Nordenskiöld wrote as though Darwinism had been merely a passing phase ([1929] 1946, 528, 616).

The modern Darwin industry has its origins in the studies made to commemorate the centenary of the *Origin* in 1959 (Churchill 1982). By this time attitudes had changed; natural selection was now the most popular mechanism of evolution, and Darwin could be regarded as the originator of both the general idea of evolution *and* of its most plausible explanation. Eminent biologists such as Sir Gavin De Beer, Sydney Smith, and Ernst Mayr began to take an interest in the development of Darwin's ideas. De Beer and Smith in particular focused attention on the enormous Darwin archive, now housed in the Cambridge University Library. The sheer richness of the material contained in his private papers tempted an increasing number of historians to explore Darwin's life and career. The discovery of natural selection seemed to represent the key step in the emergence of modern evolutionism, and it would have been embarrassing to admit that an earlier generation of biologists had not found the theory very plausible. Darwin had become a hero of modern science because his theory was the one that mattered today, and the role played by non-Darwinian theories was quietly forgotten.

The assumption that Darwinism became the dominant force in late-nineteenth-century evolutionism thus arose in part because it satisfied the values of biologists who were certain that natural selection is the most fruitful explanation of evolution. Simon Schaffer (1986) argued that the emergence of modern science was marked by a tendency to identify great heroes who were responsible for the discovery of unambiguous truths about nature. Clearly Darwin has become such a hero, and the myth surrounding his name is designed to promote the view that natural selection is the most important mechanism of evolution—a genuine discovery that will never be replaced. Like most hero-myths, it is a distortion of a much more complex state of affairs, a story manufactured to focus attention on what scientists regard as a major step toward present-day understanding. In such a story there is no room for rival theories; if they exist at all, they must, by definition, be false, and so they can be ignored as blind alleys that lead away from the main line of scientific advance. This mythicizing technique is apparent in Loren Eiseley's *Darwin's Century* (1958). Here the pre-Darwinian history of evolutionism is manipulated to show that earlier naturalists were really groping their way toward Darwinism. The emergence of Darwin's theory is, of course, the main event. Although Eiseley admitted that some biologists opposed natural selection, he concentrated on the problems that would eventually be solved by Mendelian genetics. Modern Darwinism became a kind of jigsaw puzzle; Darwin had already put together most of the pieces, leaving just a few to be inserted by later biologists.

History is not an objective factual chronicle of past events; it is an interpretation of the past by people whose perception is shaped by their position in the present. All cultures have myths about their origins, designed to legitimize their members' assumption of a privileged status. The scientific community is no exception, and Darwin has become the hero or founding father in the creation-myth of modern evolutionism. The myth is formed by distorting history for a particular purpose, a technique already familiar in the political arena. There each party or ideology seeks to influence the general view of the past in a way that will vindicate its own values. Modern historians call this technique "Whig history," after the political ancestors of the Liberal party in Britain who rewrote the country's history to emphasize how its social development had been shaped by their own principles (Butterfield 1931). "Whig history" now refers to any historical account told from the viewpoint of those in power. Such an account distorts the past by picking out a main line of development whose inevitable end product will be the triumph of those now telling the story. Any event that does not fit into the scheme is either ignored or distorted so that it does fit. This is exactly the pattern followed by the scientific community to create the conventional image of the Darwinian Revolution.

That the biologists themselves should promote such an interpretation is hardly surprising. From the perspective of a scientist, the emergence of the currently successful theory is bound to represent the most important part of the story. And no one doubts that searching for the origins of modern theories is a legitimate part of the history of science. But professional historians of science are taught to look behind the facade of official mythologies for a glimpse of the more complex reality. Many traditional stories of scientific discovery have already been exploded or shown to be based on one-sided interpretations. Yet the myth of the Darwinian Revolution has remained largely unchallenged. Historians have developed a more complex image of Darwin without questioning the assumption that his work must be seen as the key to the emergence of modern evolutionism. The explanation for this shortsightedness lies further afield, because the myth of the Darwinian Revolution serves a function both in science itself and in the broader perception of the role played by science in the development of modern thought.

In his analysis of the religious debates set off by the *Origin*, James R. Moore (1979) noted the extent to which the attitude toward Darwin's role has been shaped by the presumption of a constant state of war between science and theology. The warfare image was extensively promoted in the late nineteenth century by J. W. Draper (1874) and A. D. White (1896), both of whom saw science as a force driving back the superstition and ignorance encouraged by traditional religion. The appearance of evolutionism has always been viewed as an important step in the anticipated triumph of science's rationalist and largely materialist view of the world. Moore shows how generations of historians have thus been misled into producing highly polarized accounts of the Darwinian debate. This approach does not encourage a sympathetic account of evolutionary theories that were less radical than Darwin's, since these theories represent a compromise position in what was supposed to be a confrontation between totally opposed value systems. Those who welcome science's "triumph" over religion — including a fair proportion of modern Darwinists — would naturally prefer to believe that the victory was won in a single dramatic battle initiated by the *Origin of Species*.

But what of writers opposed to Darwinian "materialism"? Here, too, the myth of the Darwinian Revolution has a role to play. Robert Bannister (1979) argued that the image of late-nineteenth-century thought dominated by "social Darwinism" was conjured up by liberal historians anxious to show how far the modern world had advanced beyond such brutality. The link with Darwinism was a useful way of emphasizing the amorality of unrestrained capitalism, but of course its effectiveness rested upon the assumption that Darwinism had triumphed in Darwin's own time. Other writers take a more pessimistic view, since they feel that modern life is still degraded by the materialism that Darwin's theory is thought to symbolize. These critics generally concede that Darwinism has gained a dominant position in Western thought, but they do not agree among themselves on the best tactic to adopt in dealing with the historical origins of this situation.

One critical approach at least suggests some revision of traditional priorities in history. From the time of Samuel Butler (1879), some opponents of Darwinism have argued that Darwin himself should be deprived of his place as the founder of evolutionism. Treating Darwinism as a dead end, they insist that the true source of evolutionary thinking should be sought in earlier writers such as Lamarck. At the height of the *Origin* centenary, C. D. Darlington campaigned against the "myth" that gave Darwin the central role in the growth of evolutionism (1959, chap. 12), while H. G. Cannon (1959) revived Lamarck's claims. More recently Søren Løvtrup (1987) dismissed Darwinism itself as a myth that should be demolished. He, too, presented Lamarck as the true founder of modern evolution and insisted that the hollowness of Darwin's theory has always been apparent to anyone free to think clearly about it. According to this interpretation, Darwinism only gained a hold because it fitted the ideology of capitalism and the materialist prejudices of many scientists. Here, born out of sheer distrust of Darwinism, is the counter-myth of Lamarck as the source of the only effective approach to evolution (Barthélemy-Madaule 1982).

The exponents of this countermyth do not deny the power and influence of Darwinism. They merely lament this state of affairs and hope that a return to their own values will allow Lamarck's priority to be recognized. Other opponents of Darwinism use history as a means to emphasize the theory's success, since this allows them to present it as a dogmatic orthodoxy that has never allowed any challenge to its authority. The Lamarckian alternative, if recognized at all, is depicted as a lost cause trampled upon by the Darwinians' intolerant materialism. Thus Arthur Koestler's (1971) account of the early-twentieth-century Lamarckian Paul Kammerer presented him as an isolated figure driven to suicide by the already well entrenched synthesis of Darwinism and genetics. Despite being highly critical of Darwinism, Koestler did not realize that Lamarckism once exerted a considerable influence on biology. Nor did he suspect that the early geneticists at first rejected the possibility of an alliance with Darwinism. In his anxiety to portray Kammerer as a martyr to materialist persecution, Koestler positively welcomed the traditional assumption that Darwinism had already gained a stranglehold on science by the end of the nineteenth century.

A number of more responsible historians share Koestler's distaste for the implications of Darwinism and seem to have been similarly misled into exaggerating the theory's success. Both Jacques Barzun

([1941] 1958) and Gertrude Himmelfarb (1959) portrayed Darwin as an exponent of what they regard as the increasingly dominant materialism of nineteenth-century thought. John C. Greene (1959) saw Darwin as a somewhat aberrant product of the materialist tradition stemming from Descartes's cosmology. None of these writers is willing to admit that anti-Darwinian (and hence sometimes nonmaterialist) ideas may have played an important part in the growth of evolutionism. Unwilling to concede a role for the alternatives to natural selection, they are forced to believe that the success of evolutionism means the success of Darwinian materialism. Such an interpretation has its attractions for those who see modern thought as the bitter fruit grown in soil prepared by the triumph of materialism. Darwin's theory is used to symbolize modern materialism, and Darwinism's victory in the nineteenth century is proclaimed so that the moral chaos of our own century can be seen as a direct product of the materialism thus unleashed. To admit that much evolutionary thought has been non-Darwinian in character would undermine this neat explanation of the modern predicament. Historians who distrust Darwinism have thus been led to accept the modern Darwinians' own self-centered assumption of their theory's early success. Both sides promote the conventional image of the Darwinian Revolution because it serves their very different purposes in concealing the extent to which non-Darwinian forces have shaped the growth of evolutionism.

Having explained the origins of the conventional image of the Darwinian Revolution, I must now develop my alternative history of evolutionism. Here Darwin plays an important, but not pivotal, role; the developmental tradition emerges as the driving force of nineteenth-century evolutionary thought. In theory it ought to be possible to write a history of the field in which Darwin is not the central figure. But my present purpose is destructive — a clearing of the decks for action so that the possibility of an alternative approach can be more widely appreciated. I must thus continue to focus on Darwin in a significant part of my exposition in order to present the arguments against the orthodox historiography. I suggest, however, that even when thinking about Darwin's role, it is necessary to ask new questions about how his ideas were related to the continuous thread of developmentalism that unifies so large a part of nineteenth-century thought.

Chapter Two

DARWIN'S ORIGINALITY

✿ At one time it was fashionable to depreciate Darwin's intellectual achievements. Even scientists who were themselves Darwinians tended to think of him as a patient compiler of facts who merely did his best to make sense out of what he saw. They believed that the theory of natural selection would have been more or less self-evident to anyone who had undertaken the right kind of research program. Darwin's position in history, therefore, rested on his being the first to ask the right question, not on the originality of his answer. Others had dared to challenge creationism, but he was the first to see the need for a complete theory of divergent evolution based on the adaptation of species to their environment. Since there appear to be only two conceivable mechanisms of adaptive evolution, then — given that Lamarckism had already been proposed and discredited — it was only a matter of time before someone came up with the alternative of natural selection.

If scientists tended to see the theory of natural selection as the obvious product of any serious research into the problem of evolution, the opponents of Darwinism had a very different reason for minimizing the amount of intellectual effort involved. They sneered at Darwin's lack of philosophical sophistication and his inability to appreciate the weaknesses of his theory. It is still often claimed that he was led to develop the selection theory because he had absorbed the competitive social ethos of the time. Natural selection was no more than a projection onto the living world of values derived from Victorian capitalism. Whatever the scientific factors that put Darwin in a position to create such an image of nature, his originality lay only in the fact that he was the first to draw upon a source of inspiration that would soon become commonplace.

Both of these efforts to diminish the originality of Darwin's thinking could draw upon another line of evidence to support their case — the widely held belief that several other naturalists should be counted as the co-discoverers of natural selection, the most prominent being Alfred Russel Wallace. The scientists, of course, argued that the involvement of several discoverers supported their interpretation of the theory as a self-evident induction from the facts. Any biologist in the same position as Darwin would eventually have had exactly the same idea. Sociological critics argued that the parallel application of the same idea by several different naturalists was an indication of the all-pervasive cultural pressure imposed by capitalist ideology. Everyone exposed to that cultural environment was bound to think along the same lines. Independent historians may be inclined to question the validity of an argument that can be used to support two such widely divergent interpretations of the theory's origin.

To defend my thesis that Darwin's basic theory does not express a central theme of nineteenth-century thought, I must obviously challenge both attempts to demolish his originality. If natural selection really was self-evident to any biologist willing to think about the facts, the theory ought to have conquered the scientific world as soon as it was published. Or, if the mechanism was merely an expression of capitalist ideology, scientific and social Darwinism must be inevitable products of the Victorian era. My thesis requires an interpretation that steers between these two extremes by showing that Darwin *used* scientific facts and ideological principles but put them together in a way that to some extent transcended what was "obvious" to his contemporaries. To sustain this view I must also undermine the credibility of the co-discoverers of natural selection. Darwin must be seen as a unique figure, not one among many who were being led in the same direction by the scientific facts or the Victorian cultural environment.

Fortunately, the scholars of the modern Darwin industry have developed a much greater respect for his intellectual achievement. Far from being a mere fact-gatherer, he had to grope his way toward a workable theory through a maze of conflicting biological and social ideas in what is now seen as a classic example of truly creative thinking (Gruber 1974). In the words of David Kohn's introduction to his massive compilation of modern Darwin scholarship, "Charles Darwin was a thinker of profound intellect and influence" (Kohn 1985, 1). My thesis rests on this assessment of his originality but asks if such a degree of creativity necessarily guarantees widespread influence on contemporary thought. Is it not possible that Darwin's ideas were so radical that they stirred, but in the end failed to capture, the imagination of his age? The claim that the Darwinian theory acted as a stimulus to the development of nineteenth-century evolutionism, but did not altogether control the direction of that development, requires an image of Darwin as a thinker who was in touch with contemporary ideas but whose interpretation of the issues differed significantly from that adopted by the majority of his fellow scientists.

My first task must be to survey the ways in which the new Darwin scholarship has extended and, in some cases, challenged, the traditional view of how he developed his theory. It is necessary to determine how he was led to the more radical aspects of his thought which anticipate the principles of modern Darwinism. But it is essential to beware of the temptation to concentrate solely on those parts of the theory which are now considered important. In other areas, Darwin continued to work with orthodox ideas that the modern biologist sees as an obstacle to the theory's full development, especially in the area of heredity. Modern historians, realizing the importance of building up a unified view of Darwin's natural philosophy, avoid the rejection of his nongenetical approach to heredity as an embarrassing failure. By recognizing that Darwin both used and transcended contemporary ideas, it is possible to see how his theory can have been both welcomed and misunderstood. If some so-called Darwinists failed to grasp what are now regarded as the theory's most radical and fruitful concepts, this was because Darwin's own perception was still partly shaped by concepts that could easily be fitted into a more conventional view of how nature develops.

A New Biology?

The orthodox interpretation of Darwin's career reflects a number of interests. It is based in part upon Darwin's own assessment of how his ideas had developed, as presented in the autobiography written toward the end of his life. The autobiography was first published (in an abbreviated form) by his son Francis, who edited the *Life and Letters of Charles Darwin* (1887). This collection helped to structure the story of Darwin's career, which has led Jon Hodge (1985a) to write of the "Franciscan" view of the Darwinian landscape. Hodge points out that the Franciscan view remained intact until the early 1960s and was accepted in de Beer's biography (1963). This survival presumably indicates that most aspects of the Franciscan interpretation were acceptable to the scientists who founded modern Darwinism. In general, the story picked out those aspects of Darwin's work which could be seen as having permanent value for the development of evolution theory.

The basic outline of this orthodox interpretation is so well known that it need only be summarized here (for a more detailed and partly updated version, see Bowler 1984a, chap. 6). Darwin's early medical studies at Edinburgh are dismissed as having had no real impact. The story really begins when he moved to Cambridge in 1827, where he remained a dull student but began to expand his interest in natural history. Here he was captivated by William Paley's *Natural Theology* (1802), with its descriptions of how species were adapted to their way of life. Although soon to abandon Paley's explanation based on divine contrivance, Darwin remained convinced that adaptation was the most important phenomenon that a theory of evolution would have to explain. In effect, he stood Paley on his head by turning adaptation from a divinely ordained state into a natural process. This influence prevented him from becoming enmeshed in the simple-minded progressionism soon to be expounded in Chambers's *Vestiges of Creation*. Meanwhile, Darwin's extracurricular studies of natural history brought him to the attention of the professors of botany and geology John Henslow and Adam Sedgwick.

As a result of this adequate, if unofficial, training as a naturalist, Darwin was offered the opportunity that was to transform his life. He was asked to go as naturalist aboard HMS *Beagle*, a survey ship being sent out by the Admiralty to chart the waters around South America. The five-year voyage of the *Beagle* (1831–36) has captured the popular imagination as a romantic journey of exploration (Moorehead 1969; Ralling 1978). Serious historians of evolutionism pinpoint two key consequences of the trip: it turned Darwin into an enthusiastic supporter of Charles Lyell's uniformitarian geology, and it forced him to recognize the problems faced by a creationist in trying to explain the geographical distribution of species.

As he studied the rocks of South America in the light of Lyell's *Principles of Geology* (1830–33), Darwin rejected the idea of catastrophic earth movements in favor of a gradualist or uniformitarian model of change. This created a framework within which it became possible to think of gradual change among living things, instead of mass extinctions and divine creations. Lyell also gave Darwin a methodology to work with. He advocated the study of the past solely in the light of processes that could be observed in the present. Although Lyell himself was unwilling to admit it, uniformitarian geology paved the way for evolution in biology. Once the method was applied to the study of living things, a theory of natural transmutation was the only acceptable explanation of how new species replaced old.

Lyell rejected Lamarck's evolutionism but acknowledged that his theory would require an explanation of how species are distributed based on a historical process governed by the possibilities of migration and adaptation within an ever-changing environment. The voyage made Darwin aware of the extent to which geography controls distribution by enabling him to observe that each continent had its own characteristic types of animals and plants. He discovered fossils confirming that extinct South American forms were related to the present inhabitants of the continent. Such facts could only be explained by the creationist model through an arbitrary assumption that each continent was the focus for a "center of creation" that produced only certain types of species. The crucial factor that converted Darwin to evolutionism was, however, the much more restricted problem of geographical distribution posed by the Galapagos Islands. Here was a group of volcanic islands several hundreds of miles out in the Pacific Ocean, inhabited by creatures related to, but differing in significant details from, those of the South American mainland. Darwin soon discovered that each island had its own characteristic forms. The finches in particular have come to be seen as a paradigmatic example of speciation due to geographical isolation (Lack 1947). Once Darwin was sure that the finches were indeed distinct species, not merely varieties of a single species, he had to accept either that the Creator had performed a separate miracle for each island, or that the transmutation of isolated populations adapting to a new environment was capable of producing new species.

Darwin chose the latter alternative soon after his return to England, and he almost immediately accepted that evolution must be responsible for even the major developments in the history of life on the earth. As a result of his Galapagos experiences, he necessarily saw evolution as a branching, adaptive process, not as a linear progression. But how exactly did the adaptation take place? In his search for a plausible mechanism, Darwin began to study the one area in which species could actually be seen to change, namely, the realm of animal breeders. He soon realized that the varieties of pigeons, dogs, and the like were produced by artificial selection. Each population contained an assortment of individuals who differed from one another in an apparently random manner, and the breeder simply picked out those whose characters best suited his needs. By breeding only from those individuals with a particular kind of variant character, the breeder could eventually enhance and fix that character in his animals. Inspired by this model, Darwin began to search for a natural equivalent of the selective process. He found it when reading Thomas Malthus's *Essay on the Principle of Population*, which showed that the tendency of any population to breed beyond the available food supply would result in a constant "struggle for existence." Darwin realized that those individuals best adapted to the environment would do better in the struggle than their rivals and would thus survive and transmit their characters to future generations.

This insight was the basis for the theory of natural selection, which Darwin had worked out in enough detail by 1842 to produce a sketch; he wrote a more substantial essay in 1844. According to the orthodox interpretation, the 1844 essay contains the mature version of Darwin's theory. His refusal to publish it thus becomes something of a puzzle, usually explained away by noting his fear that the time was not yet ripe and his desire to build up his scientific credentials before publication. His major research program on barnacles was in fact undertaken to fulfill the latter ambition. The timing of the eventual publication of the theory was dictated solely by the appearance of Alfred Russel Wallace's paper outlining his independent discovery of the concept of natural selection.

The traditional, or Franciscan, view of how Darwin's theory was created has been modified and in some respects challenged by modern scholarship. Unfortunately, the nonspecialist may well find himself or herself daunted by the sheer quantity of the Darwin industry's output. A recent article that merely outlines the current literature is over fifty pages long (Oldroyd 1984); a major new collection of Darwinian papers (Kohn 1985) includes over four hundred pages on the development of the theory. Darwin's notebooks and private papers have now revealed the full extent of the influences affecting him and the complexity of his thought-processes. Parts of the old story have had to be abandoned or significantly revised, while debate still rages over the interpretation of some key steps in his discovery. Some authors argue that there is still room for a major reappraisal of Darwin's thought, based on a recognition of the extent to which the traditional story has concealed his involvement with issues that are no longer taken seriously today.

The new image of Darwin is more dynamic, portraying a man whose ideas continued to change long after the preliminary version of his theory was set down in 1844. He is also more in touch with contemporary developments, constantly rethinking his own position in response to the ideas of others and fully aware of the need to describe the more radical aspects of his theory in a comprehensible language. The new Darwin scholarship still recognizes that a combination of certain influences directed him toward a unique world view that included a thoroughly naturalistic approach to evolution and the striking anticipation of some modern ideas. Historians are still working toward an understanding of how Darwin came to propose such an original theory. At the same time, however, it is now increasingly recognized that the theory was constructed within a framework that included elements more characteristic of the nineteenth century than of the modern view of nature. The old historiography dismissed Darwin's ideas on heredity and variation, for instance, as "failures" that would only be corrected by the emergence of modern genetics. The Darwin industry of today is more inclined to treat the theory as a complex whole, functioning adequately within a conceptual framework that both overlapped with and transcended the values of its time. I shall argue that an appreciation of the nonmodern aspects of the theory is vital to an assessment of its position in the development of evolutionary science. Darwin's views on "generation" (reproduction) are of interest not because they reveal gaps that had to be filled in by later biologists but because they may explain why the more radical aspects of the theory were so easily misunderstood or ignored by those scientists who were converted to the basic idea of evolution by Darwin's writings.

Some of the latest scholarship helps to flesh out the skeleton provided by the traditional viewpoint. No one, I think, would seek to challenge the claim that the most innovative aspects of Darwin's thinking arose from his conversion to a Lyellian view of nature and his efforts to understand the problems of geographical distribution within this context. Modern studies have filled in the details of this process and provide a far better understanding of the intellectual environment within which Darwin was working. The results are sometimes quite surprising. Frank Sulloway (1982) has shown that the Galapagos finches cannot have played the key role often attributed to them. Darwin did not realize that the finches differed from one island to another until it was too late, and he had to reorganize his collection from memory. Yet the basic point remains intact: it may have been the mockingbirds rather than the finches that provided the major illustration, but the message was the same. Populations transported from a species' main location to isolated sites would adapt to the new environments in their own way. The geographical factor ensured that Darwin would start from a branching model of evolution in which there could be no privileged line of development.

The importance of Lyell's geology in stimulating Darwin's geographical research program cannot be underestimated (Hodge 1982). It should be noted, however, that Lyell's position in the history of geology has undergone considerable revision in recent years, and he is no longer treated merely as a stepping stone to Darwinism. Although uniformitarianism promoted a gradualist view of change that was immensely important for Darwin, Lyell extended it into a completely steady-state cosmology that denied the developmentalism so characteristic of nineteenth-century evolutionism (Hooykaas 1959, 1966; Rudwick 1971). Darwin probably went further than anyone else toward acceptance of Lyell's nonprogressionist interpretation of the history of life (Bartholomew 1976), but even he could not tolerate a world view in which the age of reptiles might return in some future environment.

The real significance of Lyell's work for Darwin came from its emphasis on the attempt to understand the present distribution of animals and plants in terms of physical changes in the recent geological past. Explaining the present through a truly historical process was not an obvious technique to most of Darwin's contemporaries. The more popular type of progressionist theory invoked a predetermined trend with a predictable outcome (Bowler 1976a and chap. 3, below). Browne (1983) pointed out that even in the field of biogeography, many botanists sought mathematical relationships that had no real explanatory power. The "philosophical naturalists" wanted to do more than merely collect information (Rehbock 1983), but they often looked for abstract patterns rather than historical explanations based on natural processes. Only a handful of British naturalists saw the implications of Lyell's position and joined Darwin in the search for realistic explanations of how species came to their present locations. Brown and Rehbock both discussed the case of Edward Forbes, who — without adopting evolution — tried to account for the distribution of British plants by postulating climatic changes in the past. Yet Forbes soon struck off in an entirely different direction, seeking an abstract pattern in the rate of creation of species based on the notion of polarity. The *only* naturalists to join Darwin in a consistent search for Lyellian explanations of distribution were Joseph Dalton Hooker and Alfred Russel Wallace. Hooker was one of Darwin's earliest converts, and Wallace independently developed his own theory of evolution.

The relationship between Darwin's and Wallace's theories is a complex one and will be discussed more thoroughly in the final section of this chapter. One of the most obvious differences between them is that Wallace had no interest in artificial selection, but Darwin consistently used animal breeding as a model to illustrate his theory. The latest research has dramatically altered the perception of the role played by artificial selection in the construction of Darwin's theory, as it has modified other aspects of the traditional story (for a survey by two of the leading contributors, see Hodge and Kohn 1985). In his autobiography, Darwin maintained that he had gone to the animal breeders in search of information, had recognized the importance of selection from their work, and then had looked for a natural equivalent. His own papers from the years 1836–37 do not support this claim, however, and some historians have even denied that the selection analogy played any significant role at all. Hodge and Kohn concluded that it was important in helping Darwin to formulate his ideas, but not at the critical point when the concept of natural selection first began to emerge.

Far from casting about at random for information that might throw light on the origin of species, it is now clear that from the start Darwin was trying out hypotheses against the facts. At this time, his thinking was still conditioned by a teleological attitude toward adaptation. He did not immediately reject the traditional belief that nature was governed by a wise and benevolent God. Instead, he sought to explain how such a God might have instituted a natural process that would maintain species in a state of perfect adaptation to a changing environment. His early ideas explored an analogy between evolution and growth: new species "budded off" from their parents and might even have a fixed life span, like an individual (Kohn 1980). Soon, however, he began to suspect that external conditions must somehow affect the process of individual growth in a way that would adapt the population to any change in its environment. This skirted along the edge of Lamarckism, and, indeed, Darwin always accepted a limited role for the inheritance of acquired characters. But his experience with the breeders suggested that the relationship between the environment and individual growth is normally more complex, producing random and not purposeful variation. This paved the way for the theory of natural selection based on the superior reproductive potential of those best adapted to the environment.

The now-obvious role played by Darwin's thinking on generation, or reproduction, led John Hodge (1985a, 1985b) to develop a major critique of the traditional historiography. Hodge argued that because the most original — and most "modern" — aspects of Darwin's theory were inspired by his biogeographical studies, historians have presented him essentially as a naturalist. His excursions into the study of reproduction, culminating in his theory of heredity by pangenesis, have been treated as minor aberrations that distracted him from his important work. Hodge suggested that Darwin should now be seen as working within a framework defined equally by natural history and the study of reproduction. Pangenesis was the product of a lifelong interest in the physiology of reproduction and growth which played a major role in shaping Darwin's thoughts. For Darwin, and for later Darwinists such as August Weismann, evolution was as much a function of growth and reproduction as it was of geography and adaptation. This insight produces a major change in the interpretation of

Darwin's place in the development of nineteenth-century biology. His evolutionism no longer represents a clear break with the eighteenth-century tradition of speculation about the origin of life, which quite explicitly took reproduction as its model (Bowler 1984a, chap. 3). Darwin may even have derived significant inspiration from his grand-father Erasmus Darwin, whose *Zoonomia* (1794–96) included a theory of "evolution" in the eighteenth-century tradition. Hodge also suggested that Darwin's studies at Edinburgh under the Lamarckian biologist Edward Grant may have had far more impact than Darwin was later prepared to admit, a view endorsed in a different context by Phillip Sloan (1985). More recently still, Sloan (1986) argued that Darwin's transition to a dynamic view of nature was precipitated when he acquired an interest in the more speculative aspects of German nature-philosophy. This finally allowed him to visualize the possibility that the vitality of living forms might transcend the limitation of the specific type to give potentially unlimited variation.

Hodge and Sloan have not, of course, suggested that Darwin's theory was nothing more than an extension of contemporary speculations about growth. These ideas provided a stimulus that interacted with the more conventional biogeographical factors to bring about his conversion to transformism. Nevertheless, it is clear that Darwin's science was not quite the forward-looking research program depicted by the Franciscan model. It drew heavily upon contemporary debates that were closely interwoven with the emergence of the developmental view of nature. The "modern" aspects of Darwin's theory emerged from and never fully transcended a world view in which any attempt to study the nature and origin of life was expected to confront the phenomena of reproduction and growth. As a result, Darwin always saw variation (the essential raw material of selection) as a function of disturbances to the process of growth. Thanks to the breeders, Darwin soon realized that such disturbances are essentially random. This is the point at which his thinking diverged from that of most contemporary evolutionists, who continued to believe that growth must somehow be predisposed to change in a particular way. But growth was still a primary factor in Darwin's view of evolution, and he could never rule out the Lamarckian possibility that an organism's development might respond to external stimuli in a positive way. Pangenesis was his attempt to understand how this vital aspect of evolution might work.

The implications of this reinterpretation of Darwin's position may go far beyond Hodge's own proposals. The most obvious alternative to the Darwinian view of evolution compares the history of life on earth directly with the growth of the individual organism, thereby encouraging the view that evolution is orderly, progressive, and teleological. Lamarckism is merely a byproduct of this fascination with growth, the most obvious symbol of a deep-rooted challenge to Darwin's image of evolution as a haphazard and open-ended process. There can be no truly historical study of life's development if it is programmed to unfold like the growth of an embryo irrespective of any changes taking place in the physical environment. Darwin's recognition that most variation is undirected allowed him to escape from the teleological implications of the growth analogy and to recognize the significance of the geographical evidence for branching evolution. Yet his continuing belief that growth is a vital aspect of evolution may explain why neither he nor his immediate followers were able or willing to destroy the conceptual framework within which the rival view that evolution is *directed* by growth could flourish. Later in the century, self-proclaimed Darwinians such as Ernst Haeckel would invoke the analogy with growth to promote an essentially progressionist view of evolution totally at variance with Darwin's branching model. Darwin himself may have been unable to identify the way in which his principles were betrayed, because he could not fault the inclusion of growth as a vital aspect of evolution.

This new interpretation of Darwin's biology must become an important factor in any attempt to evaluate the impact his theory made upon nineteenth-century thought. The developmental model of nature was preferred by those biologists whose real concern was to preserve the teleological character of the traditional view in a new guise. But it was possible for the radical implications of Darwin's own theory to be missed by biologists who saw that he had no intention of challenging the fundamental link between growth and evolution. He was thus unable to prevent naturalists whose interests were very different from his own from seizing upon the general idea of transmutation. The theory of natural selection would have limited impact on late-nineteenth-century biology, not because it was incomplete without a genetical model of heredity, but because it was conceived within a framework that offered better support to rival concepts of living development which circumvented Darwin's more radical proposals.

Although Darwin never abandoned his interest in reproduction, it is now clear that the more radical aspects of his theory underwent considerable refinement after the early version was sketched out in 1844. Most historians now accept that in his early speculations, Darwin was still a theist who hoped to produce a teleological version of evolution compatible with belief in a wise and benevolent God. In 1844 he still saw natural selection as an episodic process, occurring only when conditions changed and producing species that were perfectly adapted to the environment when it eventually stabilized (Ospovat 1981). Geographical isolation was the only factor that would allow a separate branch of a once-unified population to diverge from the original type to form a new species. Over the next decade or more, Darwin was forced to rethink his position completely. The result was a theory offering a far more serious threat to the teleological view of nature.

Darwin's extensive study of barnacles revealed a considerable amount of variation even within species that seemed to be living in a stable environment. This threatened the whole idea of the perfect adaptation of the majority of individuals. Darwin also became aware that Richard Owen and others were proposing a new interpretation of the fossil record as the history of the process of specialization of each class, with numerous branches diverging steadily further apart from the earliest, most generalized form (Ospovat 1976, 1981). Although these trends were not presented in an evolutionary light, Darwin accepted that his theory would have to explain them. Eventually he realized that a trend toward divergence would be produced by natural selection favoring those varieties and species that specialized for a particular way of life and thereby avoided competition from their neighbors. Evolution would now occur even in a stable environment, since further specialization would always give individuals an additional advantage in coping with the environment. Significantly, Owen's representation of divergence was explicitly modeled on K. E. von Baer's theory of how the embryo grows by specialization. Although wedded to a branching image of the history of life, Owen's use of the growth analogy illustrates how easily the idea of a directed trend in evolution could be exploited by those who preferred to see the process as the unfolding of a purposeful divine plan.

As a consequence of these developments, Darwin rejected his original belief that geographical isolation, as in the Galapagos Islands, was essential for the production of new species. He now began to suspect that divergence would occur more readily where related forms coexisted in the same area and were thus driven apart by the advantages to be gained from specialization. This change of emphasis was to have important consequences after the theory was published (Sulloway, 1979a). Darwin was never able to explain how a single population could be divided into separate breeding groups without a geographical barrier to prevent interbreeding that would reblend the potentially divergent characters with those of the general population. His difficulties on this score encouraged opponents to suggest alternative explanations of speciation based on saltations. Many modern Darwinists regard his abandoning of geographical isolation as a betrayal of one of his most important insights.

By the time Wallace's paper forced him to begin writing the *Origin of Species*, Darwin had already started work on what was intended to be a large-scale presentation of his revised theory (since published as Darwin 1975). The new theory was far more dynamic than the old. Instead of brief episodes of evolution which transferred species from one state of perfect adaptation to another, natural selection was now seen as a continually acting force that operated upon species whose characters were never fixed and never sharply defined. In effect, Darwin needed a new definition of the species concept that would uncouple it from the old idea of fixed morphological units. He did not formulate such a new definition, possibly as a deliberate policy aimed at minimizing the extent to which his theory would be seen to undermine traditional ideas (Beatty 1985). Modern Darwinism defines the species in terms of the breeding population and regards this as a true expression of the principles inherent in Darwin's theory. Yet Darwin's failure to confront the problem of interbreeding posed by his rejection of geographical isolation as a key step in speciation would have prevented his formulation of this definition. He continued to discuss species as though they were tolerably well defined units. Variation was seen not as the circulation of different genetical characters within the breeding population, but as a departure from the norm produced by disturbances of individual growth.

Although some aspects of Darwin's thinking were constrained by the accepted values of the time, in other respects his ideas now revealed their potential as a threat to the existing world view. To begin with, it is clear that Darwin himself did not perceive this threat. It has been argued that his theory can best be seen not as a reaction against, but as a product of the tradition represented by, Paley's *Natural Theology* (Cannon 1961). Modern scholarship has confirmed that Darwin saw the original version of his mechanism as the means whereby a benevolent God ensured that species remained perfectly adapted to a changing environment. Young (1971) pointed out that the metaphor of natural *selection* seems to have been chosen to convey the impression of a superintending power equivalent to the animal breeder. After the introduction of the principle of divergence, however, Darwin could no longer believe that adaptation was perfect, and he seems to have become more aware of the cruelty in nature that was an inevitable byproduct of so selfish a mechanism of evolution. Ospovat (1981) argued that to preserve his faith in divine providence, Darwin now began to focus on progress rather than adaptation as the ultimate purpose of evolution. While accepting that in a branching model of development there could be no single goal of progress, he nevertheless began to see natural selection as a force that would tend to produce "higher" levels of organization, at least in the long run.

The relationship between Darwinism and progressionism is a complex one. Some modern Darwinists have argued that natural selection does not generate progress in any meaningful sense of the term (e.g., Simpson 1973). Historians have debated the extent to which Darwin himself shared the Victorians' widespread faith in progress (Mandelbaum 1971, 77–88; Moore 1979, 158–60; Greene 1981, 95–157). It is clear that he was well aware of the difficulties that his theory raised for the idea of necessary or inevitable progress. It would be virtually impossible to compare the levels of organization achieved by entirely distinct branches of the tree of life. Specialization might sometimes lead to degeneration, as in the case of parasites. Most commentators have agreed, however, that Darwin sought to preserve a notion of progress that would evade these difficulties. Natural selection did not guarantee progress, but it did allow progress to occur as a frequent byproduct of the drive toward better functioning organisms. Here again, Darwin remained true to the spirit of the times.

A recent survey of Darwin's religious views concluded that his continued faith in progress must be seen as a function not of his theory's basic principles but of his own residual belief in a divine purpose (Brown 1986; see Brooke 1985). Darwin's religion is as complex a topic as his progressionism; most historians recognize that his faith gradually diminished in the course of time, but many argue that he never quite abandoned the hope of reconciling evolutionism with some inscrutable divine purpose. He recognized the extent of cruelty in nature, but he hoped that the development of higher organisms would justify the suffering involved. His early notebooks reveal that he almost immediately realized how a theory of natural evolution would imply a materialist account of human origins. The human race would become a product of nature, which would make it all the more necessary to see nature itself as a purposeful system. Darwin's reading of Malthus's *Essay on Population* was part of a deliberate program of research into topics that would throw light on the natural origin of human faculties. In his autobiography Darwin said that Malthus alerted him to the "struggle for existence" in nature, which became the agent responsible for picking out the better-adapted individuals. A major insight on the way to the construction of the theory of natural selection thus came from a work that played a central role in the social debates of the time.

In the Air

Darwin's admission that Malthus's principle of population was crucial for his recognition of the struggle for existence raises one of the most controversial issues surrounding the origin of his theory. If Darwin saw natural selection as a theory of progress through struggle, does this not suggest that his theory represents an expression of contemporary social values? Malthus's principle was a contribution to the school of political economy that hailed the competitive individualism of laissez-faire capitalism as the key to economic progress. There have always been some critics who have dismissed natural selection as a simple projection of the capitalist ethic onto nature. It has even been argued that Darwinism was somehow "in the air" at the time, as though Victorian culture was moving inexorably in this direction (Millhauser 1959, chap. 3). Capitalism demanded that the whole world should be portrayed in its own image, and natural selection represents the most obvious way of applying the idea of progress through struggle to biology. If Darwin had not made the connection, the argument goes, someone else would soon have filled the gap and sparked the general conversion to a competitive view of nature and society.

My general position obviously requires me to challenge this claim. If natural selection really was the most obvious way of applying Victorian values to nature, it would be impossible to defend the view that Darwin's theory had only a limited impact on nineteenth-century thought. Scientific and social Darwinism would both have to be seen as expressions of a cultural trend that was inevitably replacing divine miracle with free-for-all struggle as the only mechanism of creation and progress. The claim that Darwinism was in the air makes light of those aspects of the theory that threatened to undermine a simplistic form of progressionism. It glosses over any distinctions that might be recognized among the various ways in which the concept of struggle could be used to create a mechanism of change. And, of course, it ignores the fact that natural selection was not accepted as the dominant mechanism of evolution by the biologists and social scientists of the late nineteenth century.

Darwin himself had no doubt that his theory was not "in the air." In his autobiography he wrote, "I tried once or twice to explain to able men what I meant by natural selection, but signally failed" (Darwin 1887, 1:87; Darwin 1958, 124). This would certainly suggest that few others were responding to whatever cultural influences shaped Darwin's thinking. In their desire to preserve the claim that the scientific method represents a thoroughly objective way of studying nature, modern scientists have been reluctant to admit that external factors may have played a role in the creation of this, or any other, theory. In their eyes, Darwin was led to his theory by his choice of research program, and natural selection was merely the most fruitful solution to his technical problems. In his biography of Darwin, De Beer (1963) played down the role of Malthus's principle and ignored the analogy with capitalism. According to this model of scientific discovery, any use of a successful theory by political writers represents mere opportunism on their part and gives no evidence of an ideological component in the creation of the theory. A similar view of the relationship between science and society runs throughout Ernst Mayr's epic survey of the history of biology (1982).

The apparent parallel between natural selection and capitalism was noticed almost as soon as the theory was published and led critics to suggest that Darwin had merely projected the values of his own society onto nature. In 1875 Friedrich Engels noted that the theory simply transferred the doctrine of economic competition into nature and was then used to claim that competition was "natural" in human society (Marx and Engels 1953, letter to P. L. Lavrov, 176; see the *Dialectics of Nature*, quoted ibid., 186). Since then, the claim that natural selection expresses the spirit of capitalism has been made by thinkers as diverse as John Maynard Keynes (1971–79, 9:276) and J. D. Bernal (1969 4:1085). Robert Young has been one of the most eloquent modern exponents of the view that Malthus's principle and Darwinism are both expressions of a "common context" provided by the social attitudes of the time (1969; repr. in Young 1985a). In the most recent celebration of the Darwin industry's efforts (Kohn 1985), Young openly criticized the ever more minute examination of Darwin's technical biology and proclaimed that "Darwinism *is* social" (Young 1985b). Such views have inevitably been taken up by the sociologists of science, who argue that scientific knowledge is merely opinion used as a tool for social purposes (Barnes and Shapin 1979; Shapin 1982). Although proud of their emancipation from the simple economic determinism of early Marxism, the sociologists of science still seem eager to exploit such an apparently obvious congruence between an important theory and its social environment.

The Darwin industry responded to Young's charge by pointing out that without his particular scientific background, Darwin would not have been in a position to formulate his theory. But modern Darwin scholars have no intention of treating him as a figure divorced from the social and cultural environment of the time. They might well argue that Darwin was only able to complete the more radical aspects of his program because he was the first to realize that contemporary social values had begun to undermine the credibility of natural theology. It is now known that in addition to Malthus, Darwin gained insights from reading the work of Adam Smith and other political economists (Schweber 1977). The concept of divergence through specialization reflects the economic advantages supposed to accrue from the division of labor. Even the Darwinian concept of species seems to reflect the individualist model of society. The orthodox typological view of species can be compared with a totalitarian political philosophy; individual organisms must conform to the specific type just as individual humans beings are supposed to submerge themselves in the higher reality of the state. To some extent, Darwin saw the species as a population of varying individuals with no fixed type, just as *laissez-faire* economics sees society as composed of individuals with divergent interests. The possibility that Darwinism reflects certain aspects of Victorian society is accepted by the majority of modern scholars, but this does not commit them to the view that natural selection is nothing more than a projection of the capitalist ethic onto nature.

To what extent, then, was Darwin's use of the struggle for existence modeled on the Victorians' view of their own society? There seems little doubt that people were becoming increasingly aware of the role played by struggle in society and in nature. Political theories and even the general literature of the time seem to emphasize the theme of struggle in an unmistakable way (see Gale 1972 for wide-ranging survey). If carried beyond a certain point, however, recognition of struggle threatened to undermine faith in God completely, as in the often-quoted lines from Tennyson's *In Memoriam*, written between 1833 and 1850. Tennyson wrote of nature as indifferent to life itself and to the beliefs of man,

> Who trusted God was love indeed
> and love creation's final law —
> tho' Nature, red in tooth and claw
> with ravine, shriek'd against his creed.   (*In Memoriam*, st. 56)

These words can certainly be seen as an illustration of the Victorians' growing willingness to acknowledge the harsher side of nature, but they are hardly an anticipation of natural selection. The cruelty of predator to prey has little to do with the struggle that must take place *within* each species to ensure that useful changes occur.

The majority of Victorians could not accept that struggle and suffering were without purpose. Their faith in progress was an essential means of reassuring themselves that whatever the short-term suffering, there was a meaningful goal to be achieved. Although Paley saw a benevolent God creating a fixed universe in which happiness was always maximized, the Victorians rationalized their ever-changing world by supposing that progress toward better things could be gained through struggle. Progress occurred because God intended it to, but a new spirit of self-confidence encouraged the belief that the motor of progress had been built into nature. The Victorians were creating a new world through their own efforts, and they wanted to feel that the results were endorsed by the Creator of the system in which they functioned. A new system of morality emerged in which nature rewarded those individuals who contributed toward progress and penalized those who did not pull their weight. A competitive system was the most obvious way of ensuring that progressive individuals would be rewarded at the expense of the "unfit." The link with the ideology of capitalism is obvious enough. In the free-enterprise system, each individual seeks his own success, but to give the system a moral purpose it is assumed that he will only achieve his goal if he contributes to the overall good of society.

The problem with assuming that natural selection is a simple expression of this philosophy of progress through struggle is that there are two quite different ways of visualizing how the motor of change might work. Something must ensure that useful characters are enhanced within the population. Darwin modeled his suggestion on the activity of the animal breeder; the struggle for existence picks out those individuals who *by chance* possess useful characters and allows them to breed. The mechanism is strictly hereditarian, since the fate of each individual is decided by the characters it inherits from its parents. There is no point in the individual's trying to work harder or improving its habits, since however hard it tries, rivals who are congenitally better equipped will beat it in the end. This is a philosophy of trial and error, a game of genetical Russian roulette in which the chances of inheritance determine everyone's fate.

Far from being an expression of the Victorians' faith in the purposeful nature of struggle, Darwin's mechanism ignores one of its most cherished values — the role of individual effort and initiative. From the largely static social philosophy of Malthus to the evolutionary sociology of Herbert Spencer, struggle was always seen as a factor that was designed to encourage *everyone* to try harder. The elimination of the congenitally unfit was always a secondary and purely negative factor, not to be confused with the primary and positive function of stimulating effort and initiative throughout the population. Those historians who see Malthus as the source of Darwin's primary insight try to portray him as an exponent of ruthless individualism only too happy to see the poor eliminated by starvation.

The image is easy enough to sustain, since Malthus was critical of state relief for the poor. But is this interpretation valid, or does it represent a distortion intended to exaggerate the extent to which early nineteenth-century society was moving toward a Darwinian model (Bowler 1976b)?

In his effort to show that redistribution of wealth would not eliminate poverty, Malthus argued that the population has a natural tendency to grow faster than the food supply. Relief for the poor would only encourage them to produce so many children that the state would be unable to feed them even if it wanted to. Malthus realized that in primitive societies, the pressure of population would encourage a "struggle for existence" that would eliminate those tribes least able to defend themselves. Yet there is much in the *Essay on Population* that does not fit in with the claim that Malthus wished to encourage the elimination of the unfit in his own society. In the first edition of 1797 he saw the pressure of population as a divine mechanism for encouraging slothful mankind to work. The stimulus to activity was the primary concern, and Malthus was not anxious to explore the implication that those incapable of responding to the stimulus would starve. In later editions he supposed that educating the poor would help them to limit their families, thereby avoiding the need for struggle altogether. This was still part of a divine plan, since the poor would learn moral values as well as reproductive prudence (Santurri 1982). Wealth was not a prize gained by competing with one's neighbors; instead, its possession entailed a responsibility to use it for the good of society. Darwin took Malthus's image of primitive society and applied it to nature, but he was not following Malthus's view of how a civilized society should work.

Malthus justified a continuation of the existing social hierarchy within a divinely ordained universe. But in his anxiety to demolish the claims of the political radicals who called for socialism, he proposed an idea that others would use as the basis of a more dynamic view of nature and society. Charles Darwin and Herbert Spencer both saw the population principle as the driving force of change, a coincidence that has led to the bracketing of the two thinkers as the architects of the new scientific and social Darwinism. Hofstadter (1955, chap. 1) portrayed Spencer as the social Darwinist who first transmitted the new morality from Britain to America. But several more recent studies have challenged this position by offering an interpretation of Spencer's writings that distances him sharply from the Darwinian theory (Bannister 1979; Freeman 1974; Ruse 1986). Although they share some common ground, the Darwinian and Spencerian theories of evolution use the concept of struggle in very different ways. I shall return to Spencer's thought when discussing social evolutionism below (chap. 7), but to appreciate the extent of Darwin's originality it is necessary to contrast his theory with that developed by Spencer before the *Origin of Species* appeared.

Spencer's first major defense of the free enterprise system was his *Social Statics* of 1851, in which he set out to prove that individual initiative is the sole mechanism of progress (Peel 1971). He realized that one consequence of a laissez-faire system would be the elimination of the totally unfit, and he considered that this was essential lest these unfortunates hold back progress. But their removal was an essentially negative result that merely prevented the buildup of nature's occasional failures. The true source of progress was the stimulus that the struggle for existence imparted to individual activity: it forced everyone to learn how to cope with an ever-changing social environment.

> If to be ignorant were as safe as to be wise, no one would become wise. . . . Unpitying as it looks, it is best to let the foolish man suffer the appointed penalty of his foolishness. For the pain — he must bear it as well as he can: for the experience — he must treasure it up, and act more rationally in the future. To others as well as to himself will his case be a warning. And by multiplication of such warnings, there cannot fail to be generated in all men a caution corresponding to the danger to be shunned. (Spencer 1851, 378–79)

The unfit individual is not congenitally unfitted to his environment; he is merely temporarily out of phase with it. His suffering is not a prelude to extermination, but an encouragement to improve himself and hence to become fitter. This is not a Darwinian philosophy but is more closely linked with the popular enthusiasm for "self-help" summed up in Samuel Smiles's book of that title (1859).

Spencer saw nature as a stern schoolmaster who expects us to learn from our mistakes. Our children, too, will learn, until eventually there will be no one left out of step with social evolution. The Lamarckian implications of this are evident. The fact that Spencer foresaw the eventual elimination of bad habits confirms that the benefits of experience were somehow incorporated into the race. Spencer was an early convert to biological evolution and accepted the inheritance of acquired characters as the mechanism of change (see his 1852 essay, "The Development Hypothesis," repr. in Spencer 1883, 1:381–87). He did not think of natural selection until he read the *Origin of Species* but presented Lamarckism and natural selection as the joint mechanisms of evolution in his *Principles of Biology* (1864). It was here that he coined the phrase "survival of the fittest" (1:444), which has perhaps led some historians to overemphasize his interest in selection. He never abandoned Lamarckism, however, and his *Factors of Organic Evolution* (1887) argued that this effect became steadily more important as evolution progressed. Spencer's philosophy thus linked free enterprise with the analogy between growth and evolution. The characters contributing to fitness which were acquired through struggle would be inherited and would shape the growth of future generations. It was Spencer who popularized the term *evolution* in its modern context — a term that, significantly, was originally used to denote the growth of the individual toward maturity (Bowler 1975).

The combination of Lamarckism and progress was attractive to many late-nineteenth-century thinkers, including those who did not agree that struggle was the best teacher. For obvious reasons, though, Spencer's *laissez-faire* philosophy is often seen as the social parallel of Darwinism, the best illustration that selection had become an obvious mechanism of change, waiting only for a naturalist to apply it to biology. As I have suggested, the association is rather more tenuous. The two men drew upon a similar inspiration in two very different ways. Darwin never rejected Lamarckism completely, but his own insight used struggle to explain why those who have by chance inherited a fitter makeup will survive and reproduce. Far from being a simple expression of social values that were "in the air," natural selection would come as a surprise even to thinkers such as Spencer who were deeply concerned with biological and social evolution.

The Co-discoverers

One last objection must be disposed of to establish the case for treating Darwin's theory as an untypical manifestation of nineteenth-century evolutionism. Most accounts of the history of biology accept that his theory was not unique, for the obvious reason that a number of other naturalists independently discovered the principle of natural selection. The best-known of these co-discoverers is, of course, Alfred Russel Wallace, whose paper of 1858 forced Darwin into the belated publication of the theory. The claim that natural selection was discovered by several naturalists at about the same time is often used to support the view that the mechanism was an obvious way of extending capitalist ideology into science. It implies that there was indeed a cultural pressure forcing everyone to think along similar lines. Whatever the gulf between Darwinism and Spencerianism, the former must therefore be treated as an equally obvious extension of Victorian social values.

To defend my basic thesis I shall challenge the whole notion of multiple independent discoverers of natural selection. There are four serious contenders for the title: William Charles Wells, Patrick Matthew, Edward Blyth, and Wallace himself (relevant passages from their works are collected in McKinney 1971). Few modern scholars accept that Wells, Matthew, and Blyth developed anything more than superficial parallels to the theory of natural selection. Darwin's reaction in 1858 suggests that Wallace should be taken more seriously, but it is possible that Darwin overreacted and read more into the paper than was really there. Darwin himself thus established a myth of Wallace as a co-discoverer — a myth that would often be used by Darwin's opponents as they sought to diminish his reputation. I shall argue that no one, including Wallace, was able to develop a satisfactory theory of natural selection independently of Darwin. The scenario in which the selection theory emerges as an inevitable extension of capitalist ideology thus evaporates, leaving us free to make an impartial assessment of the impact of Darwin's theory.

It is worth noting that the first alleged account of natural selection came forty-five years before Wallace's paper, so there is no question of a striking coincidence in time. It was in 1813 that Wells read his paper "An Account of a Female of the White Race of Mankind" to the Royal Society (McKinney 1971, 21–28). Here he suggested that the human races might have been formed when groups moved into unoccupied territory where they encountered new conditions. Accidental variations within the population would produce some individuals better adapted to the conditions, who would thus tend to become the parents of a new race. There is only a faint hint of a struggle for existence in Wells's account, and Kentwood D. Wells (1973a) pointed out that there is no suggestion that the selection mechanism could produce new species among animals. This limited recognition of genetical trial and error in the formation of races can hardly be counted as a true anticipation of natural selection.

Patrick Matthew has a somewhat better claim to be regarded as Darwin's precursor. His theory of selection appeared in an appendix to his *Naval Timber and Arboriculture* of 1831 (McKinney 1971, 29–40; see Wells 1973b). Darwin himself later accepted that Matthew had developed a similar idea, but he pointed out that no one had noticed it in so obscure a location. Eiseley (1958, 129) accused Darwin of being less than fair in his dealings with Matthew. Nevertheless, it is clear that there are substantial differences between the precursor's ideas and Darwin's actual theory. Matthew invoked the periodic mass extinction of species by geological catastrophes, after which the few survivors would rapidly diversify into a whole range of new species by a process that does indeed look like the selection of better-adapted characters in the struggle for existence. But Matthew believed that the new and perfectly adapted species would remain absolutely stable over vast periods of time, thereby illustrating divine providence. At best, then, his theory anticipated the immature version of Darwin's views. Matthew himself did not regard his theory as an expression of contemporary social values. In a letter claiming priority after the publication of the *Origin*, he suggested that his original statement had been ignored because the "age was not ripe" for it, and that Darwin's theory was being misunderstood for the same reason (Wells 1973b, 256).

Edward Blyth's reference to the role of competition came in an 1835 paper entitled "An Attempt to Classify the Varieties of Animals" (McKinney 1971, 43–56). It has received more attention than it probably deserves, thanks to Eiseley's claim that Darwin picked up the idea of selection from Blyth before reading Malthus and then concealed his debt (1959 repr. in Eiseley 1979, 45–80). Subsequent research into Darwin's papers has failed to confirm such an influence, and no historian now takes Eiseley's claim seriously. The influence must have been very faint even if Darwin had noticed the reference to competition in Blyth's paper, since Blyth did not see competition as a truly constructive force. He remained true to the old view of divine providence, in which the elimination of the unfit is seen as nature's way of preserving original types.

Alfred Russel Wallace is by far the most important on the list of possible co-discoverers, since the orthodox history of evolutionism accepts that he, at least, hit upon the essence of the Darwinian mechanism. The paper he sent to Darwin in 1858 is supposed to contain a clear outline of natural selection, clear enough to have forced Darwin to begin the move toward publication. Wallace has thus been accorded a seat in the pantheon of evolutionary heroes and has merited a number of recent studies (George 1964; Beddall 1969; Williams-Ellis 1969; McKinney 1972; Fichman 1981). Some commentators argue that Wallace has never been given the recognition he deserves, implying that Darwin and his followers deliberately kept him out of the limelight. Wallace himself showed no resentment, however, and those who feel that he was slighted have to contend with the obvious fact that Wallace's discovery came twenty years after Darwin's. In recent years, the attention of those who wish to raise Wallace's reputation at the expense of Darwin's has focused on the principle of divergent evolution. There have been two claims (Brackman 1980; Brooks 1984) that Darwin's recognition of divergence was inspired by one of Wallace's earlier papers (1855), although this claim is repudiated by the

Darwin industry (Kohn 1981; for a detailed discussion of the relationship between the two men, see Kottler 1985).

As a follower of Lyell's historical approach to biogeography, Wallace certainly worked out a theory of adaptive evolution. A number of recent studies, however, have suggested that Darwin may have been mistaken when he assumed that Wallace's 1858 paper contained a theory of natural selection identical to his own. Two significant differences between Darwin's and Wallace's mechanisms have been identified, both of which, incidentally, undermine the possibility that Wallace was responding to the ethos of competitive individualism. The 1858 paper must be studied without presupposing that he was presenting a theory identical to that of Darwin. There are ambiguities in the paper, and the natural temptation has been to accept Darwin's interpretation. But if we forget Darwin and simply try to understand Wallace's own words, there are grounds for suggesting that his view of natural selection differed considerably from that normally taken as the chief mechanism of evolution. Wallace did not acknowledge any such differences in later life, but it is understandable that he would have been reluctant to admit that his original idea did not coincide with the interpretation of his paper that Darwin had popularized while Wallace himself was out of the country.

A. J. Nicholson (1960) drew a distinction between what he called the competitive and the environmentalist versions of selection and argued that Darwin's theory centers on the former but that Wallace was concerned chiefly with the latter. In competitive selection the struggle between individuals is the driving force of evolution; the less fit are always eliminated by their fitter brethren in any environment. Even in a stable environment there will be differences in fitness which will cause selective elimination and hence continued evolution. The stimulus for environmental selection is the environment itself, which sets an absolute standard that each individual must meet or die. Each organism struggles against the environment, and its fate will be decided independently of what happens to the others. Evolution only occurs when the environment changes the standard against which the species is measured. Competitive selection is thus by far the more dynamic and ruthless process, and it is clear that Darwin's concept of selection was competitive in form, at least by the 1850s. But Nicholson was able to point out a number of passages in Wallace's 1858 paper indicating that selection was being presented in the environmentalist, not the competitive, mode. Wallace thought that it would only be at times of particular environmental stress that less well adapted forms would be exterminated. Kottler (1985, 374) accepted Nicholson's claims and suggested that Wallace tended to think in environmentalist terms throughout his career. Wallace never felt comfortable with Darwin's theory of sexual selection, which is very much a product of the competitive viewpoint.

The second problem with Wallace's initial conception of natural selection centers on his use of the term *variety*. The title of the 1858 paper is "On the Tendency of Varieties to Depart Indefinitely from the Original Type," and throughout the paper Wallace writes in terms of the extermination of the less well adapted *varieties*. Darwin obviously assumed that Wallace was using the term to denote individual variants within a population. The traditional interpretation of Wallace's paper has followed this reading, thereby accepting a close parallel between Darwin's and Wallace's views of the level at which selection works. But in the mid-nineteenth century "variety" was more often used to denote what would now be called a subspecies, that is, a distinct population within a species, all the individuals of which have certain characters distinguishing them from the rest of the species. The American paleontologist Henry Fairfield Osborn first suggested (1894, 245) that Wallace's 1858 paper was really depicting selection at work between subspecies rather than individuals. Without being aware of Osborn's brief reference to this point, I myself developed a similar theme some years ago (Bowler 1976c). The same point was also accepted by the noted biologist Sir Alister Hardy (1984, 66). Selection that acts to eliminate well-established varieties (subspecies) is certainly part of the Darwinian theory, but it is not the basic mechanism of natural selection that acts upon individual variations. If this analysis of the 1858 paper is accepted, Wallace again emerges as someone who developed a theory of natural selection different from and less significant than Darwin's.

For the most part, Wallace simply assumed that a species will divide itself into varieties (subspecies). His main concern was to argue that the less well adapted varieties are eliminated during times of environmental stress, after which the surviving variety divides once again so that the process can be repeated. Did Wallace realize that the selection of individual variants is the primary mechanism of change, responsible for the division of the species into varieties? There are some passages in his paper that can be interpreted in this way, but even if Wallace recognized the point, it was not his main interest. The bulk of the paper outlined what is by Darwinian standards a secondary level of selection and showed no desire on Wallace's part to explore the implications of a theory of individual selection. Significantly, Wallace was not in touch with animal breeders (who of course use individual selection) and remained suspicious of Darwin's analogy with artificial selection. He had approached the topic solely through biogeography, and thus he thought in terms of selection that acts on well-marked varieties rather than individuals. Kottler (1985, 379) accepted that Wallace did not seem to be clear on the level of selection he wished to bring out in 1858. Kottler pointed out that when the paper was reprinted in Wallace (1870), section headings and footnotes were added to increase the impression that "varieties" are to be understood as individuals. These additions suggest that, at the very least, Wallace sensed the need to clear up ambiguities in his original discussion.

Both of the arguments outlined above diminish the value of the parallel between natural selection and Victorian capitalism. Although Wallace acknowledged the influence of Malthus's population principle, he seems to have used it in a very different way from Darwin. According to Nicholson's argument, Wallace did not think of competition between members of the species; he saw the species itself as competing against a hostile and limiting environment. This is an entirely plausible interpretation of Malthus, but it reveals the lack of a necessary link between the population principle and the idea of individual competition. Wallace realized that limitations on the food supply would eliminate the unfit, but he assumed that it would be unfit *groups* that would be exterminated, not unfit individuals. This would be in accord with the fact that Malthus himself discussed the struggle for existence in terms of competing tribal groups. These arguments destroy the logic of the claim that natural selection was an obvious application of *laissez-faire* economics to the natural world. Wallace (who in any case spent years away from civilized society) did not at first see nature as a scene of unrestrained individual struggle and can hardly have been responding to the ethos of capitalism. The notion that Darwinism was in the air cannot be reinforced by the assumption that if Darwin had not published the *Origin*, Wallace, or someone else, was waiting in the wings to take his place.

The *Origin of Species* was a unique book that introduced an entirely new and — to Darwin's contemporaries — an entirely unexpected approach to the question of biological evolution. Even if we accept that Wallace's 1858 paper had the potential to be worked up into a comprehensive theory of evolution by natural selection, it would have taken Wallace years to develop that potential on his own. The paper by itself would have had little effect. Indeed, even when published alongside extracts from Darwin's own writings, it aroused no great interest. It was the *Origin of Species*, not the joint publication of the Darwin-Wallace papers by the Linnean Society, that began the great debate. Had Darwin not been there to respond to Wallace's half-discovery, it is most unlikely that anyone would have taken the latter seriously. Without Darwin there would have been no theory of natural selection to disturb the scientific world of the 1860s and hence, perhaps, no major debate on the general question of evolution.

Chapter Three

THE IMPACT OF THE *ORIGIN*

✿ The term *Darwinian Revolution* came into common use because it was clear to everyone at the time that the *Origin of Species* had ignited the debate that converted the scientific world, and everyone else, to evolutionism. A study by Hull, Tessner, and Diamond (1978) confirmed that approximately three-quarters of the British scientific community accepted evolution during the years 1859–1870, and Elegård's (1958) survey of the periodical press revealed a widespread popular conversion in the same period. Ellegård also stressed, though, the extent to which popular evolutionism was non-Darwinian in character. This points to a theme I shall develop at length below, namely, the extremely limited influence of Darwin's theory of *how* evolution worked. Some of his most vocal supporters had little real enthusiasm for natural selection, and positively anti-Darwinian theories flourished in the later decades of the century. Darwin converted the scientific world to evolutionism, but not to Darwinian evolutionism, even though some biologists proclaimed themselves to be "Darwinians." All too often, their Darwinism turns out to be little more than pseudo-Darwinism, a blend of Darwinian rhetoric with attitudes that can more fruitfully be seen as continuations of the pre-Darwinian view of nature. As the next chapter will show, it *is* possible to identify differences between pseudo-Darwinians and anti-Darwinians, but they are by no means as clear-cut as one may at first imagine.

This situation leads to an apparent paradox. If Darwin's theory was not widely accepted, why should it have been the vehicle that converted scientists to the general idea of evolution? Why should biologists who had little real interest in the Darwinian approach have been so impressed with the *Origin* that they wanted to call themselves Darwinians? The answers to these questions lie in the character of

47

pre-Darwinian biology. It would be wrong to say that evolutionism was in the air, at least within the British scientific community, but it is possible to identify certain trends within the orthodox biological tradition that were making a simple-minded creationism seem ever more implausible. The idea that nature is a dynamic system designed to unfold through time in a purposeful manner had many attractions. But this idea also violated certain highly respected elements of the traditional world view. These elements would have to be abandoned if a developmental view of the history of life was to become plausible. The reluctance of naturalists to face up to this intellectual surgery created a situation in which a sharp external stimulus was necessary. The *Origin*, by proposing a detailed argument for evolutionism backed up by a radically naturalistic explanation of how the process might work, forced everyone to accept the challenge. Some younger biologists made a rapid readjustment and were only too glad to acknowledge Darwin's influence even if they did not accept the whole Darwinian message. More conservative thinkers openly set out to formulate anti-Darwinian theories that would prevent the takeover of the idea of evolution by what they regarded as a thoroughgoing materialism.

If many late nineteenth-century versions of evolutionism were pseudo-Darwinian or anti-Darwinian in character, both containing elements salvaged from an earlier view of nature, then historians should acknowledge the need for a new picture of how the evolutionary viewpoint emerged. Instead of treating Darwin as the central figure, he will have to be seen as a radical who gained an eminent position because his theory served as a catalyst that helped other biologists to come to grips with their own very different problems. From this perspective, Darwin can be compared to a lumberjack who frees the logs jammed in a stream but has no control over the direction in which the stream is flowing. There is a whole tradition of conceptual development to be explored, a tradition that is non-Darwinian in character and that merely underwent a period of metamorphosis in response to the stimulus of the *Origin*. The pre-Darwinian biologists who thought about the development of life on the earth should no longer be dismissed as secondary figures. Instead they should be seen as the founders of a movement that continued into the post-Darwinian world by successfully evading or opposing the challenge presented by Darwin's theory. There is a direct continuity linking Robert Chambers's *Vestiges of Creation* (1844) to the use of the recapitulation theory by the "Darwinian" Ernst Haeckel and the "anti-Darwinian" neo-Lamarckians. Any analysis of the post-Darwinian era must avoid the temptation to assume that anti-Darwinian ideas are irrelevant or to exaggerate the significance of issues that can, with hindsight, be seen as paving the way for the emergence of modern evolutionism.

Of course there were some genuine Darwinians in the late nineteenth century, and their work will be described in a later chapter. But it is now time for the story of the developmental view of nature to be told in its own terms, with the impact of Darwinism viewed as a useful intrusion by an idea whose full potential could not be recognized by the majority of its contemporaries. New questions must now be asked. How far had the developmental approach moved toward a theory of the transmutation of species before Darwin came on the scene? What were the problems that caused the move toward a non-Darwinian view of evolution to become blocked and, hence, created the need for an outside stimulus? Was Darwin's ability to influence this alien tradition due partly to the skill with which he and his early followers played the game of scientific politics? In a later chapter I shall have to investigate how pseudo-Darwinians such as Haeckel preserved the essence of the developmental viewpoint and ask why Darwin failed to challenge the perversion of his theory by some of his self-proclaimed followers. Perhaps the anti-Darwinian evolutionists of the 1890s will have to be seen as the true heirs of the developmental tradition, since they exposed its non-Darwinian character more clearly than those who expressed loyalty to the Darwinian symbol. Instead of presenting the "Darwinian Revolution" of the 1860s chiefly as the starting point of twentieth-century Darwinism, it will be necessary to accept that in terms of its own cultural environment, it was merely an episode in an ongoing process within a tradition that would seem quite alien to most of today's evolutionists. The source of Darwinism's transition from a secondary to a primary role in biology will then have to be sought in a later episode, in which the developmental viewpoint was destroyed by forces stemming only in part from the original Darwinian challenge.

The Gestation of Nature

The French philosopher and historian Michel Foucault (1970) argued that eighteenth-century thinkers could not have anticipated the modern idea of evolution because they saw the realm of natural objects as a rationally ordered system. Foucault suggested that it was the comparative anatomist, Georges Cuvier, who established the open-ended view of natural relationships that made Darwinism possible, even though Cuvier himself rejected evolution. Since Darwin supposed that the existing species are the products of a haphazard process of migration and adaptation, his theory was certainly incompatible with the view that species can be linked together by an artificially structured pattern. But how successful was Cuvier in convincing the majority of nineteenth-century naturalists that they should abandon the belief that nature is a rationally ordered system? Cuvier had destroyed the old idea of a linear pattern of natural relationships—the chain of being (Lovejoy 1936). Historians have tended to assume that once this was gone, the only alternative was a more flexible view of how nature is organized. But a surprising number of nineteenth-century biologists continued the search for an underlying pattern in nature, more complex than the chain of being but with an equally rigid structure. Their ideas on the "shape" of the pattern dictated their view of how life develops on the earth, since the history of life was still seen as the preordained unfolding of a rationally ordered plan.

The belief that nature is built to a rational pattern drew much of its inspiration from the morphological approach to the study of living things. Morphology is the science of form, the attempt to describe the structure of things and to understand the laws that govern the way a certain class of objects is structured. Much of nineteenth-century biology was still essentially descriptive and, hence, morphological in character. It involved the study of the structure of living organisms and sought to explain how the different forms were related to one another. It would be misleading to suppose that the study of form was conducted without any interest in the physiological processes that built and maintained the living body, but all too often it proved far easier to describe form than to understand function. In their efforts to go beyond mere description, the morphologists tried to understand the laws governing the construction of living bodies, but these laws were often conceived in a purely morphological sense. They were not laws of cause and effect as understood by the materialist, but patterns that were seen as limits imposed upon the diversity of natural objects. It was hoped that the bewildering variety of living forms could be unified by recognizing their underlying pattern.

Embryology was still a largely descriptive science, and morphologists were fascinated by the pattern of development revealed by the growing organism. One of the most pervasive sources of a non-Darwinian approach to evolutionism was the belief that the pattern of individual growth is the key that allows the biologist to understand the history of life on the earth. In the words of Robert Chambers, science would reveal "the universal gestation of nature" (Hodge 1972). The recapitulation theory, based on the assumption that the growing embryo repeats the steps in the history of life revealed by the fossil record, became popular long before evolutionism but went on to become a major feature of evolutionary biology (Russell 1916; Gould 1977b). The analogy between growth and evolution was non-Darwinian in character because it encouraged the belief that evolution shares the progressive and teleological character of individual growth. Just as the embryo grows inevitably toward its mature form, so the history of life ascends through a fixed hierarchy of stages toward its goal. Darwin himself shared the belief that reproduction and growth are essential aspects of the evolutionary process, but he escaped the fascination with directed patterns of evolution because he recognized that variations in the growth pattern are produced at random and then selected by the environment. Many nineteenth-century biologists assumed that variation is not a disturbance of, but an *addition to*, growth, constrained to follow a course already marked out by the existing stages of development. The belief that the evolution of life has been produced by a series of additions to growth upheld the recapitulation theory and allowed both pseudo- and anti-Darwinian biologists to deny that evolution is a largely haphazard process.

The analogy with growth promoted a developmental, or genetical, rather than a truly historical view of the past. By accepting that evolution is directed by adaptation rather than the existing growth-process of the individual, Darwin was forced to adopt a genuinely historical approach to explaining how species came to be formed and located as they are observed today. The hazards of migration and adaptation ensure that each species has been shaped by a unique sequence of natural events, not by a preordained sequence of stages through which its development must unfold. The study of geographical distribution, coupled with the idea that the earth's geography has changed through time, creates an impression of evolution as a process solving a bewildering variety of ecological problems on a largely *ad hoc* basis. The unity of nature lies not in the pattern of development, but in the constancy of the physical laws governing all aspects of the process, however complex. Evolution becomes a haphazard, branching, and open-ended process, with no rigid trends and no goal toward which everything is striving.

The growth analogy, by contrast, upholds the view that evolution must be orderly, goal-directed, and, hence, in a very real sense, the unfolding of a preordained pattern. It thereby preserves exactly those features of the creationist view of nature that Darwin challenged. Whether one believes that the pattern unfolds according to a divine plan, or as the result of potentialities built into the fundamental character of nature, the aim is still to ensure that development is teleological in the sense that it is directed toward a particular goal.

All nineteenth-century naturalists realized that the history of life has been a very complex process, but many of them preserved the belief that there is a central theme running through the variety of natural developments toward a single goal. Although he called himself a Darwinian, Haeckel used the recapitulation theory to present an image of the tree of life as a structure with a "trunk" that runs upward to the human race as the pinnacle of natural development. Haeckel thus evaded the lesson taught by K. E. von Baer's much earlier demonstration that the embryo grows by a process of specialization, not by the ascent of a hierarchy defined by a series of "lower" forms. Von Baer, like Cuvier, had tried to destroy the notion of a linear hierarchy of forms with the human race at the top, offering instead a branching model of relationships far more compatible with Darwinism. But his teaching was constantly evaded by those biologists who remained convinced that some aspects of growth must represent the ascent of a linear pattern. Even those naturalists who modeled their view of the history of life explicitly on von Baer's divergent image of growth still tended to stress that the process had a built-in direction imposed by the steady addition of more specialized characters. The theory of branching evolution could thus be given a non-Darwinian flavor through the assumption that the appearance of specialized variants on the central theme occurs in a preordained manner.

The developmental view of nature had obvious potential as the foundation for a non-Darwinian view of evolution. Once the growth of the embryo was taken as the model for the pattern in which life develops on the earth, the implication that the latter process might also unfold through gradual transformation could hardly be avoided. Instead of a planned sequence of miraculous creations, surely the Designer could have programmed nature to reveal itself according to a preconceived pattern in which each form spontaneously transmutes itself into the next highest stage in the plan through some appropriate stimulus. Such ideas did indeed emerge some time before the *Origin of Species* was published, and they remained popular in one form or another long afterward. But there were certain aspects of the traditional view of nature that hindered the movement of the developmental viewpoint toward an evolutionary interpretation of its basic philosophy. Conservative thinkers in particular objected to the extension of the system to include the human race. If mankind were just the last product of a progression through the animal kingdom, the supposedly distinct character of human moral faculties would be threatened. Even more pervasive among naturalists was the feeling that species must be treated as fixed entities, distinct units in the plan of creation. How could one species be transformed into another without the destruction of the fixed character of specific types? One solution was to insist that transmutation is an episodic rather than a continuous process — that the new species bud off almost instantaneously from the old by an evolutionary saltation or jump. In the pre-Darwinian world, however, there was a good deal of reluctance to speculate on how this process might work, if only because the search for a "mechanism" of change might undermine the belief that the sequence of developments is preordained.

Early nineteenth-century naturalists were aware that mechanisms of transmutation had been proposed, most obviously Lamarck's theory of the inheritance of acquired characters. The traditional historiography of the Darwinian Revolution has always been based on the assumption that such pre-Darwinian ideas were completely discredited. To support this model, it was assumed that critics such as Cuvier and Lyell had been able to ensure that Lamarck was largely ignored. It is now becoming evident that this is only a half-truth, arising from a concentration on the more conservative naturalists, especially in the British scientific community. Radical thinkers in most European countries *were* taking transmutation seriously as part of their challenge to the status quo. Even in Britain, the scientific naturalism associated with the era of the *Origin of Species* was anticipated in the 1830s by political agitators and medical people drawn from the rising middle classes. From the start, evolutionism was perceived as a vehicle for promoting a more mobile view of social relationships. These early initiatives were successfully resisted, however, allowing a more conservative attitude to remain dominant in science during the 1840s and 1850s.

Continental Europe was the source of the more radical thinking that promoted pre-Darwinian evolutionism. Even in France, where Darwinism would be largely ignored, transformist ideas were openly discussed in the early decades of the century. Historians have traditionally assumed that Cuvier's influence was sufficient to block any general acceptance of Lamarck's views. It has now been shown that Cuvier did not enjoy absolute power within the French scientific community (Outram 1984). Rivals such as Geoffroy Saint Hilaire were able to encourage significant support for evolutionary ideas (Bourdier 1969). Pietro Corsi (1978) has shown that Lyell's critique of Lamarck in his *Principles of Geology* was almost certainly inspired by a realization that the theory was still popular in France.

In Germany, too, a more dynamic view of nature was emerging in the early part of the century. Germany was the center from which many new developments in biology radiated out to the rest of the world. Morphology had flourished under Goethe and the idealist school of thought known as *Naturphilosophie*. It was hoped that the study of form would reveal an underlying pattern to confirm that nature was the expression of a rational Mind. Long after the speculative excesses of *Naturphilosophen* such as Lorenz Oken had been repudiated, the search for an underlying pattern in nature continued. Timothy Lenoir (1982) introduced English-speaking historians to the complexities of German biological thought in this period, pointing out that a research program known as "teleomechanism" arose in opposition to speculative *Naturphilosophie*. Although mechanistic in principle, this school retained a commitment to teleology and encouraged a great deal of morphological research into the structure of the growing embryo. It was within this program that J. F. Meckel and others first suggested that the development of life on the earth may parallel the sequence of morphological stages seen in the growth of the human embryo.

But was the idea that life on earth develops through a teleological pattern of growth necessarily linked to transmutation? As I shall describe below, the Swiss naturalist, Louis Agassiz, certainly did not think so. August Weismann, recalling the attitude of his teachers during the 1850s, declared that they had avoided the whole idea of transmutation because it smacked too strongly of *Naturphilosophie's* excessive speculation (Weismann 1904, 1:28). Yet there were certainly *some* German naturalists who supported evolution (Temkin 1959; see Mayr 1982, 389). Von Baer himself was prepared to see some aspects of the development of life in evolutionary terms, although his respect for teleology ensured that he would repudiate the logic of Darwin's theory (Lenoir 1982, chap. 6). Weismann's experience suggests that German naturalists, too, had become stymied by a conceptual logjam. The developmental view of nature made creationism look ever more implausible, but those who tried to initiate an explicitly evolutionary approach tended to indulge in wild speculations that merely put the majority off the whole subject. The analogy between growth and the development of life on earth created a framework within which it would be possible to formulate a developmental theory of evolution, but the analogy could not by itself induce naturalists to confront the conceptual problems that would have to be resolved (see fig. 3). German scientists were ready for evolutionism, but some new initiative would be needed to make them take the subject seriously.

The traditional image of pre-Darwinian British natural history is of a largely descriptive enterprise, apart from the obligatory references to divine providence inspired by Paley. The work of Adrian Desmond (1984, 1987, 1989) has now revealed the extent to which this con-

Figure 3. Pre-Darwinian view of life's development on earth, from H. G. Bronn (1858). Without supporting transmutation, Bronn presented an image of continuous development that seems to anticipate the later evolutionary trees. Note that, despite the many divergent branches, Bronn included a "main stem" that moves upward to mankind.

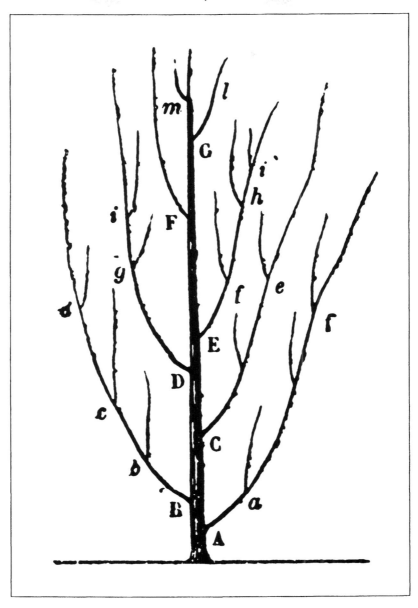

servative approach was challenged by radical thinkers during the 1830s. Anatomists such as Robert E. Grant were prepared to invoke Lamarck's theory as part of their campaign for social and professional reform. Transformism made a serious bid for scientific influence at this time and was bitterly resisted by conservative thinkers seeking to retain the idea of a divinely created universe. The anatomist Richard Owen rose to prominence as the leader of the conservative faction that successfully blocked this first onslaught of transformism, thereby helping to build up the pressures that would be released by the publication of Darwin's theory. Fear of transformism during the 1840s and 1850s was fueled by the certain knowledge that it was a theory with radical philosophical and social implications.

The problem was that even the conservatives recognized the need for something more sophisticated than Paley's argument from design. Whether radical or conservative, the 1830s was the age of the "philosophical naturalists" (Rehbock 1983). They were philosophical because they looked for underlying generalizations that would explain why nature is organized in the way we observe. This could be done in several different ways. One was through the study of biogeography, in an attempt to work out the laws governing the distribution of species. This route led Darwin and Wallace to their theories of adaptive, divergent evolution. But the majority of naturalists wanted not haphazard divergence but all-embracing regularities that would confirm their feeling that the world is the product of a rational plan of creation. It was only because Darwin escaped this fascination with orderly patterns of development that he was able to conceive a purely naturalistic mechanism of evolution.

The patterns were not all of a kind that would generate an evolutionary implication. William Sharp MacLeay's circular or quinary system of classification, for instance, supposed that the animal kingdom is divided into five basic types, which can be arranged into a circular pattern (Rehbock 1983, chap. 1). Each type consists of five classes, also in a circle, and each class contains five orders, and so on down through the hierarchy of taxonomic ranks to the species (five in a genus). Such an abstract and regular pattern of creation need have no developmental implications, although its sheer regularity fascinated Chambers, who devoted a whole chapter of *Vestiges of Creation* to it (1844, 236–76). The implications of such a program are obvious: if nature really could be shown to be built upon such an artificially structured pattern, it would be clear that the species were not the products of a purely haphazard process of development. Design by a supernatural agent would be inescapable.

The transcendentalism of the Scots anatomist Robert Knox offered a less rigidly structured way of seeing an underlying pattern in nature (Knox 1862; Rehbock 1983, chap. 2). Drawing on the idealist philosophy popular in Germany, Knox saw all species as modifications of a basic archetypal form. The "unity of type" that Darwin was to explain as a consequence of common descent was seen as an indication that nature is built to a rational plan. It was Richard Owen who went on to become the greatest exponent of this approach within the British scientific community. His *On the Nature of Limbs* (1849) argued that the underlying unity of type among the vertebrates is *better* evidence of design than mere adaptation. Although Paley treated each species individually, the transcendental anatomist tried to find orderly arrangements linking the species that could only be explained as the product of a rational Architect of nature. This is what I have called the idealist version of the argument from design (Bowler 1977a).

Owen and many other comparative anatomists had a deep interest in the bewildering variety of extinct species now being revealed by the fossil record. Inevitably, they sought to use the notion of a divine plan to explain the pattern of development in the history of life. The most obvious underlying pattern was a progressive development toward higher forms, which had become apparent by the 1830s (Bowler 1976a). The earliest rocks yielded only the remains of invertebrates, with the first fish appearing in the Silurian. The Mesozoic was dominated by the reptiles, including the dinosaurs. Although present in small numbers in the Mesozoic, the mammals only became dominant in the Cenozoic, gradually progressing to the more advanced creatures of today, including the human species. The vast majority of naturalists saw this sequence as a progression from the lowest to the highest forms of life and sought to explain why creation should have followed such a pattern. There were some early efforts to explain the sequence in terms of changing physical conditions; one view was that as the earth had cooled from its originally molten state, the steadily improving environment had allowed the creation of ever higher forms of life. The most influential progressionism of the mid-nineteenth century, however, sought to explain the trend not as a response to external changes but as a divine plan with a symbolic purpose, indicating the goal toward which the history of life was aimed.

The leading exponent of this new progressionism was the Swiss naturalist Louis Agassiz, who was later to become one of the founding fathers of American science (Lurie 1960). Agassiz was well aware that the overall development of life on the earth was a complex affair, but he was convinced that a central theme runs through the progressive development of the vertebrate classes, pointing toward mankind as the highest form of life. The pattern is, in fact, God's way of indicating humanity's position at the head of His creation.

> The history of the earth proclaims its Creator. It tells us that the object and term of creation is man. He is announced in nature from the first appearance of organized beings; and each important modification in the whole series of these beings is a step toward the definitive term of the development of organic life. (Agassiz 1842, 399).

The various side-branches in the history of life are merely variations on the central theme, like a composer's elaborations on a piece of music to enhance its artistic effect. The central pattern can be observed repeating itself in miniature in the growth of the human embryo, the parallelism again illustrating the orderliness of creation. Agassiz still believed that the human embryo passes successively through stages in which it is a fish, a reptile, and only finally a mammal (a view already discredited by von Baer). He thus laid the foundations of the recapitulation theory, in which the history of a type is thought to be repeated in the growth of the modern embryo (Gould 1977b). For all the efforts of von Baer and Darwin to show that growth and development represent a process of specialization, not the ascent of a linear hierarchy, this essentially progressionist and teleological image was to bedevil evolutionism through the rest of the century.

The comparison with the growth of the embryo shows how easy it would be for the new progressionism to generate a belief that there is a more or less continuous trend running through the history of life. Yet Agassiz would not put an evolutionary interpretation upon his theory. He insisted that the ascent through the classes in the history of life is discontinuous, a step-by-step process that is incompatible with any form of transmutation. He could, quite legitimately, point to the discontinuity of the ascent revealed by the fossil record, in which new forms often seemed to appear quite abruptly, with no sign of evolutionary antecedents. Darwin was worried by this point, and he had to invoke the imperfection of the record to explain the apparent gaps in the history of life. But Agassiz was almost certainly motivated by more than mere objectivity; he was actively opposed to the whole idea of continuous development. Mayr (1959a, repr. in Mayr 1976, 251–76) explained Agassiz's opposition to evolution as a result of his adherence to a typological view of species. As a good idealist, Agassiz saw each species as a form with a transcendental significance, of which the physical bodies of its living representatives are mere transient

copies. He was thus predisposed to see the history of life as a discontinuous process in which the species are distinct units arranged
into an orderly pattern of development. This may be an oversimplification of Agassiz's position (Winsor 1979), but I suspect that Mayr
has been generally correct in identifying the typological view of species as a barrier holding back the growth of an evolutionary interpretation of progressionism. The idealist naturalists of the midcentury
wanted unifying trends in the history of life, but they also wanted
fixed species, and they were unwilling to postulate evolution if it
violated the latter concept. There *was* a way out of this dilemma
through the idea of saltatory or discontinuous transmutation, but as
yet many progressionists were unwilling to admit the need for such
a compromise.

Agassiz's creationism was reinforced by his espousal of an extreme form of geological catastrophism. If the earth were periodically
devastated by great cataclysms, miraculous creation would be needed
to repopulate the planet after the resulting mass extinctions. Catastrophism has usually been seen as a philosophy of nature intrinsically
hostile to evolutionism, while Charles Lyell's alternative uniformitarian system is presented as a step toward Darwinism (Gillispie
1959). Recent work in the history of geology has considerably modified
this assessment, however (Hooykaas 1959; Cannon 1960; Rudwick
1971). Many catastrophists were more concerned with promoting a
directional view of the earth's history, based especially on the idea
that the planet is cooling down from an originally molten state. Their
belief in the violence of past geological activity was only a byproduct
of their commitment to a developmental world view. Lyell's uniformitarianism, though it certainly stressed the gradualist character
of geological change, was linked to a steady-state cosmology that
hardly any other nineteenth-century thinker could accept. Agassiz's
extreme catastrophism was thus an anomaly. By the 1840s the majority of directionalists had scaled down the catastrophes they postulated to the extent that the theory did not stand as an inevitable
obstacle to evolutionism in biology. The way was clear for a synthesis
of the directionalist and gradualist positions through the concept of
progressive evolution.

Significantly, it was an outsider to the scientific community who
first promoted such a synthesis. Robert Chambers was an Edinburgh
writer and publisher with an interest in natural history who felt that
the professional scientists were refusing to admit the direction in
which their detailed research was leading them. He wrote his *Vestiges
of the Natural History of Creation* (1844) to argue that the fossil
record indicates a gradual, progressive development of life on the

earth, best explained as a sequence of transmutations whereby species are steadily raised to higher levels of the chain of being. Traditionally, Chambers has been dismissed as someone who came "just before Darwin" (Millhauser 1959). Careful study of his book reveals, however, that it contains nothing like an anticipation of the Darwinian theory (Hodge 1972). It is more fruitful to treat *Vestiges* as the clearest expression in the English language of the developmental view of nature already gaining ground in Germany. Whatever the reaction of the scientists at the time, Chambers's theory outlined many aspects of the pseudo- and non-Darwinian evolutionary ideas that were to flourish after the *Origin* had convinced everyone that the basic idea of evolution was acceptable.

Chambers began his book with an outline of the "nebular hypothesis," which supposed that the solar system was formed from a swirling cloud of dust under the inexorable control of natural law. Historians have shown that the nebular hypothesis played a vital role in convincing many mid-nineteenth-century thinkers that nature is inherently progressive (Numbers 1977; Brush 1987; Secord, in press). For Chambers, the inevitability of this physical evolution set the scene for biological evolutionism. He was interested in progress, not adaptation, and visualized a step-by-step ascent of a linear hierarchy toward mankind. In later editions of his book he imagined a whole series of evolutionary lines advancing in parallel through the scale. Species are still distinct entities, since the transmutation events are saltatory, one species giving birth instantaneously to the next highest in the scale. Closely related species are similar *not* because they share a common ancestor but because they have independently reached the same point in the scale of organization. Changing conditions may trigger the transitions but cannot control the direction of change. The principal "mechanism" is the addition of new stages to the embryonic growth of the individual organisms; evolution is literally a process by which life on earth grows slowly toward maturity — the "universal gestation of nature." The growth of the modern embryo recapitulates the history of life on the earth because it follows the path that actually defines the stages by which life advances.

Chambers presented his theory as a contribution to natural theology, which makes an explicit appeal to teleology. The pattern of development by which individual growth is extended to mount the evolutionary scale constitutes a divine plan programmed into nature by its Creator. Chambers used Charles Babbage's early computer to draw an analogy with the modern concept of a program that controls a whole sequence of operations, making God the Great Programmer of evolution. Later exponents of the analogy between growth and evolution tended to avoid any explicit appeal to a preordained plan of development, and in this respect Darwin's challenge to the argument from design was successful. But in the next chapter I shall discuss how easily the analogy was used to smuggle in an unconscious assumption that evolution must be progressive. Chambers's highly structured image of lines evolving in parallel through a fixed hierarchy of stages was also to reappear in many later non-Darwinian theories.

Chambers aimed his book directly at the reading public, bypassing the scientific community. Predictably, the scientists reacted by condemning his work in the most violent terms (Gillispie 1959, chap. 6; Millhauser 1959). Transformism, even when seen as the unfolding of the divine plan, was still unacceptable to the very conservative scientific establishment. Yet there is some evidence that the hostility was generated chiefly by Chambers's deliberate exposition of the most critical implication of evolutionism — the link it implied between mankind and the lower animals. He openly explored the moral consequences of the belief that the human species has reached its position at the head of creation by a simple expansion of the brain due to progressive evolution. This undermined the traditional faith that human beings are gifted with an immortal soul, since such an entity could never have evolved gradually from animals possessing no spiritual capacity whatsoever. For the time being, most religious thinkers were unable to appreciate the possibility of a new interpretation of mankind's position at the head of nature. Their conviction that human beings must stand completely above nature prevented them from accepting the idea that humanity may be the divinely intended goal of the universal progression.

This point emerged clearly in the writings of Chambers's bitterest critic, the Scots stonemason-turned-geologist Hugh Miller. In his *Footprints of the Creator*, first published in 1847, Miller admitted that, by itself, the notion of creation by law was an acceptable, even a plausible, extension of scientific thought. Noting that the individual organism is produced by a process of natural development, he went on to ask:

> Has *Nature*, during the vast geologic periods, been pregnant, in a like manner, with the human race? and is the species, like the individual, an effect of progressive development? The assertors of the revised hypothesis of Maillet and Lamarck reply in the affirmative. Nor, be it remarked is there positive atheism involved in the belief. God might as certainly have *originated* the species by a law of development, as he *maintains* it by a law of development; — the ex-

istence of the First Great Cause is as perfectly compatible with the one scheme as with the other. (Miller 1851, 13).

As a follower of Agassiz, Miller could appreciate the possibility that all nature may unfold through a progressive development modeled on the growth of the embryo. But he then insisted that the idea was unacceptable to the Christian because it would link mankind with the animals and destroy the concept of the immortal soul. In rejecting the idea of human evolution, Miller felt it necessary to reject the whole theory of the progressive evolution of life.

The scientific attacks on *Vestiges* were probably counterproductive, since they only encouraged people to read the book. Chambers knew the audience he wished to address, and his work thus helped to define the conceptual framework within which Darwin's very different theory would later be assessed. By 1850 the concept of progressive evolution was well known, even if officially discredited. In the decade leading up to the publication of the *Origin* there was a gradual softening of the opposition to transmutation and the appearance of open support for the idea that creation took place by a lawlike process. The respected Oxford mathematician and philosopher Baden Powell wrote his *Essay on the Spirit of the Inductive Philosophy* (1855) in part to argue that the best evidence of God's power being exerted in nature is the constancy of law, not the violation of laws by arbitrary miracles. He noted that there are apparently lawlike regularities in the development of living things through time. Without openly advocating transmutation, Powell thus indirectly supported the position advanced by Chambers.

Even Richard Owen now adopted a more positive attitude, despite his earlier fears that transmutation would imply materialism. Because of his bitter opposition to Darwinism in later years, Owen has frequently been dismissed as a complete opponent of evolutionism, but it is clear that he had begun to move toward a developmental form of transmutation some time before the *Origin* was published (Richards 1987). Owen was certainly a "philosophical" naturalist, and he was hardly likely to have sympathized with simple-minded creationism. He believed in an orderly divine plan of creation, which he was now increasingly willing to see as the basis for a lawlike, rather than a miraculous, succession of forms in the course of the earth's history (1849, 89). Unlike many contemporary naturalists, he was not completely unsympathetic to *Vestiges*, although he preferred to believe that the developmental model should be based on the divergent specializations postulated in K. E. von Baer's embryological theory (Ospovat 1976). In the years before the *Origin*, however, Owen was reluctant to specify the details of how the law of creation might work; at best he would only drop hints implying a saltative process by which divergent potentials in the original living forms are unpacked. Only in the post-Darwinian era would he openly support transmutation, claiming that this was what he had meant all along and insisting that Darwinian materialism could never account for the orderly process of living development.

However vague Owen's hints, it can hardly have escaped the notice of most contemporary naturalists that nonmiraculous creation was likely to involve some form of transmutation which would produce new species from those already in existence. The fact that no one challenged Owen's evasive stance suggests that there was a preference for leaving a veil of obscurity over the issue. Many wanted "creation by law," but few cared to join Chambers in specifying how the law would work. Conservative naturalists were content with the vague notion that the divine plan will unfold as directed by its Author. To inquire more actively into the mechanism by which the process operates would threaten the belief that divine forethought is the real driving force.

But what of younger, more radical biologists? They were already becoming suspicious of the traditional appeal to divine providence as an explanation of the purposeful character of natural development. Creation by law was too obviously a compromise with natural theology. As yet, however, they, too, were unwilling to begin an active search for a more naturalistic (or materialist) theory of evolution. This negative attitude can be seen in the writings of the young Thomas Henry Huxley, later to become one of Darwin's most active supporters. In 1854 Huxley published a review of *Vestiges*, in which he condemned the book for the vagueness of its basic thesis. It offered no scientifically testable theory, only the blind assertion that the pattern of development had been built into the universe by its Creator. Such a position was of no use to a scientist looking for a mechanism of transmutation that would be susceptible to empirical study. As Huxley later noted, the resulting sense of frustration was so great that he and his colleagues were inclined to say to creationists and transformists, "A plague on both your houses" (Darwin 1887, 2:196–97). This was exactly the attitude among the German naturalists of the 1850s which Weismann recalled.

Huxley's reaction to Darwin's theory shows that he was prepared to accept a naturalistic mechanism as the basis for a scientific study of living origins. Yet Huxley himself made no effort to look for a mechanism; he was content to wait and see what might turn up. With the exception of Wallace's work, there is little evidence of a deliberate effort during the 1850s to tackle the problems that Darwin had already solved. Radical naturalists seem to have thought themselves into a stalemate, while conservatives actively preferred the vagueness of creation by law. It was outside science that more positive moves were being made to formulate an evolutionary view of nature and mankind. Herbert Spencer had published his first article supporting transmutation in 1851 (repr. in Spencer 1883, 1:381–87), and his *Principles of Psychology* (1855) took the gradual development of human faculties for granted. In his 1857 essay, "Progress: its Law and Cause" (repr. in Spencer 1883, 1:1–60), Spencer clearly outlined his whole philosophy of progressive evolution. Here was the basis for a new understanding of mankind's place in nature that would address the fears expressed by Miller and other Christians by asserting boldly that mankind's cosmic significance arises from its position at the cutting edge of natural progress. Human mental and moral powers are the inevitable products of the evolutionary expansion of intelligence and awareness. In this respect the new evolutionism would have to be materialist, since it would have to treat mankind as part of material nature and the higher human faculties as merely an extension of those possessed by animal ancestors.

Even many conservative thinkers were eventually forced to abandon their faith in the distinct character of the human soul and accept the evolutionary view of human origins. Once brought to the fence, they found it easier to jump than they had feared. Spencer might talk of natural progress rather than the divine plan, but it was possible to argue that such a directed process is intended to reach its goal by the Creator of the universe that sustains it. But was the threat of an alternative philosophical system enough to force the majority into so drastic a rethinking of the traditional position on human origins? As long as the idea of biological evolution was ignored by the scientific community, it was all too easy to dismiss the radicals as a nuisance rather than to accept them as a source of new ideas that might transform orthodox values. Spencer's new philosophy of evolutionism raised issues so deep that it could be taken seriously only if the scientific community could be persuaded to declare its support for the transmutation of species. The pressure to rethink would then become inescapable.

Without the intervention of Darwin, could Spencer have persuaded radical naturalists such as Huxley to give him the scientific support he needed? They shared an essentially empiricist philosophy, so they both required a hypothesis that would explain the appearance of new species through observable processes. Spencer was convinced that such a hypothesis was already available. The inheritance of acquired characters would explain the details of adaptive evolution within the general framework of progressionism derived from his philosophy of nature. Yet in his support of Lamarckism, Spencer seems to have been a voice crying in the wilderness. Even Chambers could not take this mechanism seriously as the basis for a complete theory of evolution. In later editions of *Vestiges* he admitted that the inheritance of acquired characters might account for the purely adaptive side of evolution (1846, 235–36; 1860, 160–61), but he was interested in progress, not adaptation. Lamarckism was inadequate because it gave the adaptive principle "too much to do" (1860, 161). Chambers thought that progress had to be a distinct force in natural development, not a mere byproduct of adaptation. Spencer's great conceptual advance was to realize that Lamarckism could be the mechanism of both adaptive and progressive evolution, since a higher level of intelligence is, in the long run, the best way of coping with changes in the environment. He no longer had to rely on the vague notion of a divine plan to ensure progress, since he had convinced himself that the basic process of adaptation functions within a system whose overall structure guarantees eventual progress. His problem was to convince the scientists that Lamarckism could really work as a mechanism of adaptive evolution.

Could Spencer have eventually persuaded the biologists that they should take another look at Lamarckism? It is clear that the inheritance of acquired characters came to play a substantial role in post-Darwinian evolutionary thought, often in conjunction with the idea of progress. It is thus not unreasonable to ask whether the theory could have served as the basis for the general transition to evolutionism if the whole situation had not been changed by the publication of the *Origin of Species*. I have suggested elsewhere that Lamarckism did not have the potential to break up the logjam that had blocked both radical and conservative naturalists (Bowler 1985). In a world without Darwin, Spencer would no doubt have written his *Principles of Biology* as an integral part of his overall philosophy, and, even in its existing form, this book contains strong support for Lamarckism. In other words, Spencer would certainly have tried to get the scientists interested in his views on biological evolution in the course of the 1860s. Perhaps he would eventually have succeeded, but it is difficult to see *Principles of Biology* as the basis for a "Spencerian Revolution" within the science. Huxley, for one, remained adamantly opposed to use-inheritance throughout his career. He later argued that Lyell's refutation had been so convincing that no one was prepared to take Lamarck seriously (Darwin 1887, 2:189). Samuel Butler wrote that Lamarck had been "so systematically laughed at that it amounts to little less than philosophical suicide for anyone to stand up on his behalf" (1879, 61). Without Huxley's support, it is doubtful that Spencer could have carried the scientific community with him. Earlier in the century, Lyell and other critics had convinced all but the more radical thinkers that Lamarckism was not in the running, and it is difficult to believe that purely philosophical arguments would have persuaded naturalists to give it a second chance.

This, then, was the logjam. Spencer had created a new philosophy of progressive evolution that might force everyone to rethink traditional values, if only the scientists made transmutation an inescapable part of intellectual discourse. But his abstract approach to the question was not likely to convince the more practically minded scientists. Most naturalists accepted that there were gradual trends in the history of life that might best be explained by some form of nonmiraculous development. But the conservatives among them were evasive, preferring not to look beyond the vague notion of creation by law for fear of violating the traditional beliefs in distinct species, divine providence, and human uniqueness. Radicals wanted a more naturalistic approach, but they were convinced that the only available mechanism of adaptive evolution was unacceptable. Lamarckism became popular in Britain only much later in the century, as the search for a scientific alternative to Darwinism intensified. The situation in 1859 was clearly one in which a major new initiative in the field of biological evolution was bound to arouse intense interest. If Huxley and a significant proportion of the scientists could be forced to take evolution seriously, the obstruction would be broken up and all the implications of progressionism would have to be faced. The *Origin of Species* provided the initiative in science. It offered a new mechanism of change backed up by a skillful presentation of the evidence in favor of a theory of common descent. Both conservative and radical scientists might now be forced to think more carefully about the process responsible for the production of new species.

The Politics of a Scientific Revolution

My interpretation of the history of evolutionism accepts that the *Origin of Species* did play a key role in bringing about a scientific revolution but portrays that revolution in a new light. Instead of suddenly forcing the scientific community to accept his own approach to adaptive evolution, Darwin merely set off a flurry of activity among biologists who had encountered a mental block along the path leading to a very different concept of how life has developed. The revolution was real enough, but it was largely non-Darwinian in character, and it was spread over a considerable period of time because the actual debate over the idea of evolution formed only the central part of the process. The next chapter will reveal the extent to which some of the most popular aspects of late-nineteenth-century evolutionism can be seen as continuations of the developmental approach. Whether pseudo-Darwinian or anti-Darwinian, these theories remained true to the idea that the history of life on the earth is best visualized as an extension of the process by which the embryo grows toward maturity. Even some of the more radical naturalists who regarded themselves as Darwin's followers seem to have paid little attention to the actual model of evolution he presented. For them, the *Origin* was just a steppingstone that helped them cross over to full acceptance of transformation. Once convinced that evolution did occur, they turned their backs on Darwin's message and got on with the job of formulating their own theories of how the process worked.

If this interpretation is accepted, it becomes necessary to ask a new kind of question about Darwin's role in the transition. Since he was addressing a scientific community whose values were so different from his own, how was he able to get his message across so as to gain even the qualified kind of support necessary to get the transition underway? Is it possible that, under slightly different circumstances, Darwin's theory might have been either ignored or dismissed so quickly that it would have failed to have even a catalytic effect on the thought of others? The traditional interpretation of the debate denies this last possibility on the grounds that the *Origin* was too powerful a statement of the case for evolutionism for it to be ignored. A major debate was inevitable because Darwin had a solid reputation and had produced a book filled with detailed arguments. Unlike *Vestiges,* the *Origin* could not be dismissed as the ravings of a scientific incompetent. There is something to be said for this view, since Darwin certainly provided a devastating critique of simple creationism, and his book thus had the potential to convert even those who disliked the mechanism of natural selection. But potentially important works can be swept aside by indifference or hostility if their message is not adequately presented and defended. It is thus possible that Darwin's ability to argue his case, and to choose followers who would support him within a hostile scientific community, played a vital role in ensuring that his theory would have a significant impact.

Most accounts of the Darwinian Revolution deal at length with the debate ignited by the *Origin.* Classic surveys include those by Eiseley (1958), Greene (1959), and Himmelfarb (1959). More recently Peter Vorzimmer (1970) studied the biological debates in detail, while Michael Ruse (1979) has provided a more up to date account of the whole episode. David Hull (1973) has edited a useful survey of the scientists' reviews of the *Origin*. It is not my intention to compete with the existing literature by providing yet another survey of the arguments that raged around Darwin's book. I want to transcend the classic image, represented best by Eiseley's account, in which the opposition to Darwin's theory is either dismissed as a product of religious bigotry or analyzed mainly in terms of its ability to expose those gaps in his thinking which would have to be filled in by later research. I am more interested in the way other scientists used the basic idea of evolution once they were converted. Before exploring this issue, however, it is necessary to ask how Darwin was able to convey his argument for evolution to a scientific community that saw the development of nature in terms very different from his own. Why did "Darwinism" succeed *despite* the mass of objections raised against natural selection — and to what extent did it succeed only because Darwin's basic message could be distorted to fit contemporary expectations?

The event that best symbolizes the conventional view of Darwin's triumph is T. H. Huxley's defeat of Bishop "Soapy Sam" Wilberforce at the 1860 meeting of the British Association. Huxley was the knight in shining armor whose superior intellectual weapons destroyed the dragon of theological conservatism. It is now widely recognized by historians that the image of conflict is itself misleading, since many scientists had strong religious feelings and some theologians welcomed evolution (Turner 1974; Moore 1979). Even Huxley's encounter with Wilberforce has had to be reinterpreted (Lucas 1979). There were many in the audience at the British Association meeting who were by no means sure that Huxley had demolished Wilberforce's objections. Hooker, who also spoke at the meeting, may have done more to promote the Darwinian cause. The next chapter will show that Huxley is now seen more as a pseudo-Darwinian who had little real sympathy for natural selection or the Darwinian approach to the history of life. Huxley was an important figure not because he forced through the arguments for a Darwinian view of evolution but because his maneuvering behind the scenes ensured that the evolutionists who regarded Darwin as their figurehead were able to take over the British scientific community.

It was through persuasion and through success in the politics of science that Darwinism came to dominate British biology. There are some scientists today who resent the claim that skills in the area of public relations help a theory to gain acceptance. They feel that objective evidence in favor of the theory must be the dominant factor. Yet sociologists who study the acceptance of new ideas within the modern scientific community have shown that it is to some extent a social process (Gilbert and Mulkay 1984). David Hull (1978) has suggested that the image presented to the world by the supporters of a new theory may be very important, especially when there are apparently valid arguments both for and against the theory. The advantage will be gained by the side that presents its case most effectively, stressing the positive aspects of its own position and undermining the influence of its opponents. The successful group will evade objections or deflect them by making concessions that do not threaten its basic principles. Its members will present a united front, never falling out in public even when they have disagreements over how the theory should be applied. Biologists loyal to the Darwinian symbol gained the day because they employed these tactics and thereby outmaneuvered both the anti-evolutionists and those who wanted to found rival schools of evolutionism. Their PR skills were helped by the ineptness of their opponents, who were in any case handicapped by the need to rethink their position in response to the Darwinian threat.

To succeed in the game of scientific politics, Darwin had to play his cards very carefully. He had to present his theory in a way that would minimize the shock felt by more orthodox scientists. He also had to build up a nucleus of supporters who would rally to his side as soon as the theory came out into the open. This was particularly important, since Darwin's illness meant that he would have to rely heavily on these shock troops to carry the day in both public and private debates. It is now widely recognized that Darwin worked hard on the presentation of his theory to ensure that he would be able to communicate his new ideas. Edward Manier (1978, 1980) argued that he had to create a whole language of metaphors that would help to convey his insights to a scientific community that was not conditioned to think along such lines. Robert Young (1971) stressed the way in which the metaphor of selection itself seems to have been chosen to allow religious thinkers to retain their belief that nature is guided by a superintending power. John Beatty (1985) argued that Darwin deliberately refrained from formulating a new definition of species because he knew that fellow naturalists would not be ready for so drastic an assault on the traditional concept. Darwin also tried to present his theory in a way that would encourage biologists to believe that it was compatible with accepted views of the scientific method, as expounded by authorities such as William Whewell and Sir J. F. W. Herschel (Ruse 1975b, 1979). In all of these ways he was adapting his own ideas to the potentially hostile cultural values of the environment in which they would be evaluated. This necessarily increased the chances that his principles would be misunderstood by scientists committed to a very different view of nature. Perhaps it was important that they could be misinterpreted in this way, since a rigidly Darwinian viewpoint would not have made converts so easily.

Several recent commentators have stressed the importance of Darwin's skill in creating a nucleus of biologists who were prepared to accept his work as a new initiative that would require a re-evaluation of the whole issue (Ruse 1979, and, in a more exaggerated form, Gale 1982). Although the arrival of Wallace's 1858 paper forced Darwin to seek publication before he was completely ready, he had already taken what steps he could to ensure that his work would be regarded seriously. Obviously, he had to impress those such as Lyell and Hooker who were already committed to the view that geographical distribution should be explained in historical terms. Darwin undertook a prolonged negotiation with Hooker during which he carefully sustained the botanist's interest through a period of initial skepticism (Colp 1986). It was also vital for Darwin to have an open line of communication to the morphological school of thought, where his geographical arguments would not seem so relevant. This is where Huxley came in, and it is clear that Darwin had taken active steps to build up a good relationship with Huxley. As noted below, Huxley never made any real use of the selection theory, but his empiricist philosophy guaranteed that he would be prepared to support any new initiative in the field. If his support encouraged a misunderstanding of Darwin's real program, this was a small price to pay for ensuring that the basic idea of evolution would be taken seriously.

The early converts had to sustain the theory in public debates (as at the British Association meeting) and to promote the interests of evolutionism within the scientific community. Huxley in particular seems to have been an exceptionally skillful politician of science, instinctively making the right moves to encourage support and placate or isolate opponents. To avoid provoking open controversy, the Darwinians at first deliberately minimized discussion of evolutionary issues within the scientific societies (Burkhardt 1974). Much activity went on in the informal "X Club" to which both Huxley and Hooker belonged (MacLeod 1970). By carefully avoiding actions that would alienate doubters and fence-sitters, and by systematically promoting the interests of younger scientists who were willing to adopt the evolutionary approach, Huxley and the others gradually worked toward a majority favorable to their position. They gained control of the editorial process at scientific periodicals so that editors and referees became willing to accept Darwinian papers. The new journal

Nature was founded at least in part as a vehicle for spreading the Darwinian message (MacLeod 1969). The ever-expanding educational system was increasingly staffed by biologists willing to teach evolutionism. Ruse (1979, 262) noted that by the end of the 1860s, Cambridge University examination papers included questions on Darwinism.

It was probably fairly easy for the Darwinians to outwit the outright creationists, but there was a real chance that conservative exponents of the developmental viewpoint might organize their own explicitly anti-Darwinian school of evolutionism. This is exactly what happened in America, for instance, where neo-Lamarckian ideas began to flourish in the 1860s. But the British Darwinians were helped by the political ineptness of their chief rivals. Two of the leading proponents of anti-Darwinian ideas, Richard Owen and St. George Jackson Mivart, allowed themselves to be forced into positions on the margins of the scientific community. Owen had a difficult personality that alienated many potential sympathizers, and Mivart was virtually ostracized (Gruber 1960). There *was* an active anti-Darwinian research program in Britain (Desmond 1982, chap. 6), but it was unable seriously to challenge the influence of Huxley and his supporters.

By the 1870s Darwinism had emerged as a dominant force in the British scientific community, and the general public was well aware of Darwin's leading role in the general conversion to evolutionism. Even in other countries, the debate sparked by the *Origin* was seen as an important milestone that brought the idea of evolution sharply into focus. In America, it is clear that the neo-Lamarckians were prompted to develop their alternative view in response to the Darwinian challenge. But the existence of anti-Darwinian schools of evolutionism indicates the importance of gaining a more comprehensive understanding of how the general idea of transformism was exploited after 1859. Darwin's success had come at least in part because his theory had been skillfully presented to a potentially hostile audience. If many of the scientists who accepted evolution were committed to a very different way of visualizing how nature develops, their ideas on what might constitute an acceptable way of extending the theory might differ considerably from the program outlined by Darwin.

Chapter Four

EVOLUTIONISM TRIUMPHANT

✵ The alliance that engineered the Darwinian takeover of the British scientific community was an extremely loose one. It included biogeographers such as Hooker who were able to appreciate the arguments that Darwin had derived from the Lyellian approach to evolutionism. But it also included morphologists such as Huxley who, as will become apparent, had little real interest in natural selection or the geographical evidence. Huxley was, in fact, a typical pseudo-Darwinian; he acknowledged the *Origin* as a key stimulus in the conversion to evolutionism, but he used the idea in a manner significantly different from that advocated by Darwin. The situation was equally complex in other countries (Glick 1974; Kohn 1985). In America, the botanist Asa Gray supported Darwin with arguments drawn from biogeography, while the paleontologist O. C. Marsh adopted Huxley's version of pseudo-Darwinism. In Germany, the most prominent Darwinian was Ernst Haeckel, whose adherence to Darwinian principles was even looser than Huxley's. It was Haeckel who created the popular vision of progressionist evolutionism that was Darwinian in name alone. French biologists largely ignored the *Origin* and were only converted to evolutionism in later decades through a revival of interest in Lamarckism (Conry 1974; Farley 1974).

There were anti-Darwinian scientists in all countries, of course, and not all were creationists. Indeed, so great was Darwin's success in promoting the general idea of evolution that it is doubtful if more than a handful of biologists still accepted creationism as a viable option after the 1860s. In Britain, the anatomists Richard Owen and St. George Jackson Mivart stood out as the leaders in the attempt to define an approach to evolutionism that would oppose even the loosely defined form of pseudo-Darwinism. Such biologists had made

no serious effort to work out a theory of evolution in the pre-Darwinian era, and their ideas were thus clearly provoked by Darwin's
work. But they had no personal loyalty to Darwin as a figurehead and
searched actively for evidence that the evolutionary process functioned in a manner totally incompatible with natural selection. To
begin with, Owen and Mivart were outmaneuvered as the Darwinians
gained a dominant role in the scientific community. The dominance
was so complete that later critics such as Samuel Butler were able
to claim that Darwinism had become virtually a dogma. Only in the
last decade of the century did the stranglehold begin to loosen in the
face of growing support for anti-Darwinian ideas. Elsewhere, however,
anti-Darwinian schools of evolutionism had flourished almost from
the start of Darwin's ascendancy. The American school of neo-
Lamarckism was founded by the paleontologists Edward Drinker
Cope and Alpheus Hyatt in the late 1860s. Many German biologists
also remained opposed even to Haeckel's pseudo-Darwinism, and in
Germany, too, anti-Darwinian activity increased toward the end of
the century.

Historians have become increasingly aware of how difficult it is
to define what late-nineteenth-century "Darwinism" really was, even
in biology. All too often, the term was used as little more than a
synonym for evolutionism by writers who had no real interest in what
Darwin actually said. David Hull (1985) has suggested that it is impossible to define Darwinism as a coherent set of beliefs. Instead he
advocates a social definition: the Darwinians were simply those scientists who expressed loyalty to Darwin as the founder of evolutionism,
whatever their beliefs about how evolution actually worked. Hull
points out that Huxley did not refute Mivart's anti-Darwinian arguments and implies that Mivart could have been included within
the Darwinian camp had he refrained from personal criticism of Darwin. I think that Hull's attempt to cut the Gordian knot is just a
little too drastic. It has the merit of forcing us to acknowledge that
the gap between pseudo- and anti-Darwinians was much narrower
than earlier historians realized. But there *are* differences between the
ways in which Huxley and Mivart developed their evolutionism, even
if the differences are less interesting than the similarities. Mivart
might have become a Darwinian, but to stay within the fold he would
have had to give up certain avenues of thought, particularly his exploration of the possibility that evolution consists of parallel lines
driven along by nonadaptive forces. Huxley's writings occasionally
hint at similar interests, but he could not expand on the theme without challenging the basic Darwinian principle that species are related
by common descent rather than parallel evolution. Pseudo-Darwin-

ians often paid little more than lip service to the concept of adaptive evolution, but they could not openly explore the idea that many developments are nonadaptive in character. Anti-Darwinians welcomed evidence of nonadaptive evolution because it made nonsense out of the whole utilitarian approach.

The similarities that can be discerned between pseudo- and anti-Darwinism lie in the fact that both were largely expressions of the morphological approach to biology and the developmental view of the history of life. Both movements tended to picture evolution as the unfolding of orderly trends, without the element of haphazard divergence introduced by Darwin's concern for the accidents of migration. Evolution as perceived by the comparative anatomist in the dissecting room or museum of paleontology was a very different thing from the view revealed by the field naturalists' studies of how organisms live in the wild. In theory, the morphologist might admit that form was determined by function, but in practice formal relationships were all that mattered. For every anatomist such as Huxley who looked for patterns of evolution that at least did not violate Darwinian principles, there was another like Mivart who constructed orderly patterns based on the old idea of linear development. Darwin failed to complete a truly Darwinian revolution in biology because he was unable to wean the morphologists away from their fascination with abstract patterns expressing the underlying unity of living forms. Whatever the role played by Darwin's arguments in their conversion to evolutionism, the morphologists soon forgot the central message of the *Origin* and continued with only a slightly modified version of what they had been doing before the Darwinian stimulus came along.

Pseudo-Darwinians accepted, at least in principle, that the patterns within the history of life can be seen as essential products of an evolutionary process guided by adaptation and specialization. Anti-Darwinians stressed the nonadaptive character of some trends and were at first tempted to invoke such trends as evidence for supernatural guidance of evolution. The majority of biologists eventually abandoned explicit appeals to teleology, forcing the anti-Darwinians to propose naturalistic explanations for their trends based on orthogenesis (nonadaptive evolution governed by internal or genetical factors). In this respect, Darwin's assault on the argument from design was a success. But it was a hollow victory, since many other aspects of the pre-Darwinian developmental viewpoint remained unchanged. It might no longer be fashionable to picture the progressive development of life as the unfolding of a divine plan, but progressionism, with the human race as its inevitable goal, was still widely popular. The belief that the growth of the embryo provides the best model for the history of life was still used to create the impression that evolution was a far more purposeful process than anything compatible with Darwin's biogeographical technique.

The most typical expression of evolutionary morphology was the attempt to reconstruct the details of how life has developed in the course of geological time. Fossils had been the subject of intense public interest throughout the early nineteenth century, and it was inevitable that evolution would be linked to paleontology. The old quest for transcendental archetypes could easily be translated into a search for real (but hypothetical) ancestral relationships. Evolutionary genealogies were soon constructed to explain the origin of fossil and living forms, and every effort was made to discover new fossils that would confirm the existence of "missing links." By the 1870s, the reconstruction of life's ancestry had become a major scientific industry, which in turn became the layperson's image of what evolutionism is all about.

Yet Darwin was not the founder of this industry, since he did not see the detailed reconstruction of the history of life on the earth as the evolutionist's major task. He knew that anatomical and embryological similarities could be used to discover the degrees of relationship between modern forms, and he argued that these similarities indicate evolutionary linkages. A good classification would be, in essence, a genealogy. But recognizing that classification was based on common descent did not require the detailed reconstruction of the whole course of evolution. Apart from the abstract diagram in the *Origin*, Darwin's only representations of evolutionary "trees" for particular groups occur in his private papers (Darwin 1887, 2:343; Gruber 1974, 197). In response to critics who emphasized the discontinuity of the fossil record, he preferred to stress the imperfection of the record itself rather than speculate on what might be hidden in the gaps. Darwin saw evolution as a complex process whose past course would be very difficult to reconstruct from fragmentary evidence. He preferred other ways of studying how the process worked.

Against Darwin's advice, the evolutionary morphologists launched into the construction of detailed genealogies. Their favorite technique for filling in the gaps in the fossil record was an appeal to the recapitulation theory. I have already described how the belief that the growing embryo repeats the history of its species' past evolution emerged within the pre-Darwinian developmental tradition. Contrary to popular belief, Darwin himself was not an exponent of the recapitulation theory. His mechanism provided no justification for recapitulation: once the later stages of growth have been diverted into a new channel by variation, the old adult form vanishes and only the earlier stages of development remain unchanged. Earlier adult forms are preserved as stages in the growth of the modern embryo only if variation is an addition to, rather than a disturbance of, growth. The most enthusiastic exponent of the recapitulation theory as a guide to the history of life was Ernst Haeckel, who openly appealed to Lamarckism as an explanation of how new characters are added to ontogeny. Although Haeckel must be counted as a pseudo-Darwinian, the recapitulation theory was also a favorite tool of the anti-Darwinian paleontologists of the American neo-Lamarckian school. When a later generation of experimental biologists criticized the speculative nature of Darwinism, their hostility was in fact directed against the pseudo-Darwinism of evolutionary morphology, not against anything that Darwin himself had sought to promote. The anti-Darwinians, in fact, provide a better illustration of the developmental view of nature that the founders of modern genetics were reacting against.

Pseudo-Darwinism

I must now substantiate the claim that evolutionary morphology — even when given the name of Darwinism — was really a product of the pre-Darwinian developmental viewpoint. I propose to look in some detail at the work of two biologists who have traditionally been known as leading Darwinists: Huxley and Haeckel. If these two men can be shown to have had a stronger allegiance to the morphological research tradition, their views will provide a foundation on which to build an interpretation of post-Darwinian evolutionism that does not exaggerate Darwin's influence.

It has long been known that Huxley had problems with the selection theory. Much to Darwin's disappointment, he argued that the effectiveness of the mechanism would not be proved until a new species had been produced by artificial selection (Huxley [1860] 1893–94, 2:74). Originally, historians assumed that this was a minor quibble from a scientist whose commitment to naturalism ensured a real interest in Darwin's mechanism. The need for a major reinterpretation of Huxley's position was first suggested by Michael Bartholomew (1975) and has since been endorsed by Adrian Desmond (1982) and Mario Di Gregorio (1982, 1984). It now appears that Huxley was interested in selection only as a *possible* mechanism of evolution, a hypothesis that allowed the general idea of descent to become respectable. In his own work, he ignored even the basic concept of evolution for several years and only became interested in phylogenetic research under the influence of Haeckel.

The simple fact is that Huxley's biological interests were shaped by his career as a comparative anatomist and never really overlapped with Darwin's. Although he, too, made a voyage around the world in a naval vessel, HMS *Rattlesnake*, Huxley developed none of Darwin's concern for the problems of adaptation and geographical distribution. He returned to Britain still dedicated to uncovering the detailed structure of living forms and to searching for a unifying factor that would reduce the enormous variety of forms to a comprehensible order. He found this unifying principle in K. E. von Baer's embryological studies, which showed how the specialized adult organism develops from a more generalized starting point. In effect, embryology revealed the underlying type, of which the adult form was merely a superficial modification. Huxley's dedication to scientific naturalism arose from his opposition to Richard Owen's use of the archetype concept to uphold a mystical notion of the divine plan of creation. Desmond (1982) argued that Huxley saw Owen's conservative position as a factor that would keep science subservient to religion. To become independent and play its own role in modern society, science would have to free itself from such vestiges of the argument from design. Yet Huxley's application of the archetype concept was little different from Owen's in practice, and he himself sometimes fell into almost mystical language when discussing the concept.

Huxley had at first been suspicious of evolution because he associated it with the unfolding of a divine plan, and he wrote a bitterly critical review of Chamber's *Vestiges* (Huxley 1854). Darwin's naturalistic mechanism was important because it allowed transmutation to be divorced from mysticism, thereby opening up the prospect that the archetype linking a number of related forms might be a real common ancestor rather than a metaphysical abstraction. But Huxley could raise no enthusiasm for natural selection in practice, and his previous commitments prevented him from making any real use of evolution theory in the early 1860s. Bartholomew, Desmond, and Di Gregorio have all agreed that Huxley's detailed scientific work from this period shows no trace of evolutionism. Instead he stressed the "persistence of type" — the fact that some forms remain unchanged through vast periods of geological time — and implied that the evolutionary origins of the main forms of life might be lost in the depths of antiquity. Darwin's theory was plausible precisely because it did not require continual progress.

Huxley began to make use of evolutionism in 1868, when he at last conceded that the birds had evolved from the reptiles during the period covered by the fossil record. The newly discovered *Archaeopteryx* from the Jurassic rocks of Germany was soon seen as a classic vindication of this idea. From this point on, Huxley became a genuine evolutionist, although he still showed no interest in the actual mechanism of change and little enthusiasm for natural selection. In the words of Di Gregorio (1982, 398): "If Huxley underwent a 'conversion' to evolutionism, it took place in the late 1860s under the influence of Haeckel and German science." The original Darwinian conversion had been only a superficial affair. Desmond (1982, chap. 5) argued that Huxley's change of heart arose from his recognition of a pressing *social* need for a theory of gradual evolution. There was increasing political unrest in the later 1860s, and Huxley hoped that evolution might serve as a model to convince the working class that gradual progress was possible. Haeckel's explicitly progressionist version of evolution provided the stimulus that Huxley was looking for. Only later in his career did Huxley begin to re-emphasize his doubts about progress, as he came to appreciate the moral dangers inherent in Herbert Spencer's social evolutionism.

In his search for fossil evidence of evolution, Huxley was particularly interested in the discoveries of the American paleontologist Othniel C. Marsh. In the 1870s, Marsh unearthed a series of fossils which seemed to display the gradual evolution of the modern horse from earlier forms with a less specialized foot and tooth structure (see fig. 4; Bowler 1976a, 132). Huxley proclaimed this sequence "demonstrative evidence of evolution" (1877, repr. in Huxley 1893–94, 4:132). Subsequent discoveries have revealed that the true story of horse evolution was more complex than Huxley and Marsh imagined. They took a series of fossils and arranged them into a simple pattern of increasing specialization, assuming that this represented the actual course of evolution. It is now known that the evolution of the horse involved not a simple linear trend but a complex process of branching and extinction. Huxley and Marsh oversimplified because they were looking for the least complicated morphological pattern linking the known forms, and, allowing for the heat of debate, they can perhaps be forgiven for treating the linear sequence based on the first fossils as proof of evolution. Nevertheless, their assumptions illustrate the anti-Darwinian potential of the morphological approach. Marsh's great rival, the neo-Lamarckian paleontologist Edward Drinker Cope, argued that such regular trends could never have been produced by the selection of random variation. The linearity of fossil trends based on morphological patterns became the chief line of support for neo-Lamarckism and orthogenesis (see below).

Although Huxley was opposed to Lamarckism and never supported orthogenesis, it is clear that he was strongly tempted by non-Darwinian ideas. He was not very interested in adaptation and seems to have felt that evolution would to some extent be controlled by

Figure 4. Marsh's fossils from the horse family arranged to give the impression of a continuous (and hence linear) process of specialization. The illustration was widely used by evolutionists; this example is from Wallace (1889).

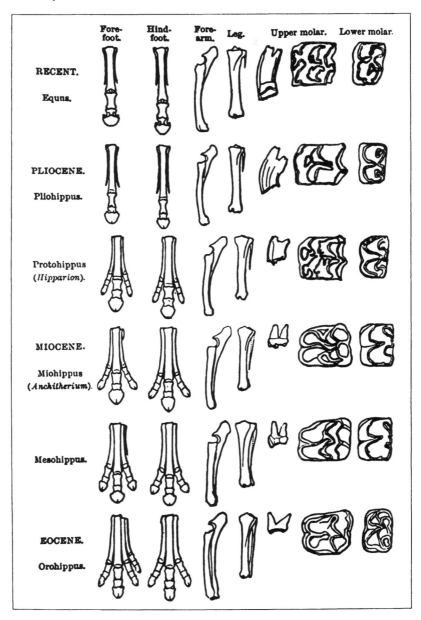

internal (what might today be called genetical) factors that would produce change whatever the state of the environment. In his review of the *Origin* he criticized Darwin's gradualism and suggested that evolution might sometimes work in a saltatory manner (repr. in Huxley 1893–94, 2:77). A paleontologist can invoke sudden leaps quite happily, since they will solve the problem of the discontinuity of the fossil record. Only the field naturalist wonders how the mutated individual will find a living and a mate when surrounded by unchanged members of the original population. In 1871 Huxley wrote of a "law of variation" that would somehow direct evolution, although he conceded that he was unable to formulate such a law (1893–94, 2:181–82). By 1878 he was openly arguing that evolution might be directed along definite lines, producing trends that would simply be weeded out by natural selection if they moved in a harmful direction (1893–94, 2:223). There were many naturalists for whom it was but a short step from this position to a thoroughgoing anti-Darwinism based on the assumed existence of nonadaptive trends.

Marsh, too, claimed to be a Darwinian, but the superficiality of his commitment is evidenced by his efforts to portray evolution as an inherently progressive force. He proposed a "law of brain growth" according to which all mammalian groups showed an increase in brain size throughout the Tertiary (Bowler 1976a, 137). He gave a loosely Darwinian explanation of this trend: "In the long struggle for existence during Tertiary times, the big brains won, then as now" (Marsh 1878, 55). Yet neither Darwin nor Huxley would have endorsed such a general progressive trend, since they were well aware that the physical specialization of many modern families has taken place without an increase in overall intelligence.

The discovery of fossil evidence for the origin of the birds and of the horse family were two of the great successes of evolutionism. They encouraged biologists to hope that other hypothetical missing links would also be confirmed by hard evidence. So great was the pressure to extend the program that it was soon being applied in areas where fossil evidence was most unlikely to be discovered. Evolutionary morphologists searched for the origins of the basic types, or phyla, of the animal kingdom, which seemed to be lost in a time long before significant fossilization began. Evidence from comparative anatomy and embryology thus became crucial. A whole industry centered on trying to understand how the first vertebrates had evolved from an invertebrate ancestor (see Russell 1916, chap. 15). The evolutionary morphologists thus embarked, often in the name of Darwinism, on a project that Darwin himself saw as highly speculative — the construction of a complete history of the development of life on earth.

One of the most important tools in the armory of evolutionary morphology was the recapitulation theory, pioneered by Haeckel and imported into Britain by the young E. Ray Lankester, who supervised the translation of Haeckel's *History of Creation*. Lankester supported Darwinism because he saw the goal of biology as the explanation of life's development in materialist terms, and in principle he realized that the history of life would be an irregularly branching tree shaped by adaptation. Later he discussed the possibility of degenerative as well as progressive evolution, pointing out that degeneration would obscure the original sequence by which the higher forms had evolved (1880). Yet his enthusiasm for the recapitulation theory inevitably led him to suggest that the development of life can be modeled, at least in principle, on the growth of the embryo. He proposed (1877) that the history of life can be seen as a process with a main line of development leading toward mankind; each stage in the main ascent throws off side branches whose status is defined by their point of origin. Despite his warnings about degeneration, this image of the "main line" was repeated in the article on zoology he wrote for the *Encyclopaedia Britannica* (Lankester 1888).

The subordination of embryology to phylogenetic research was also a central theme of the research school founded by Francis M. Balfour at Cambridge. Balfour's *Treatise of Comparative Embryology* of 1880–81 stressed the value of descriptive embryology for reconstructing the development of life on earth, which he regarded as a project founded by Darwin himself. Like Lankester, he endorsed the selection mechanism (1885, 2:2–3) while in practice paying little attention to the problem of how new characters would actually evolve. Unlike Lankester, however, Balfour's main interest centered not on the recapitulation theory but on the assumed tendency of early embryonic stages to reveal the common starting point for later divergences. He was also more aware of the extent to which the practicalities of embryonic growth would obscure the remnants of ancestral stages. As a result of these more cautious attitudes, Balfour seldom endorsed the kind of linear progressionism found so often in Lankester's work. His was a more divergent — and hence more "Darwinian" — model of evolution. Yet even Balfour's phylogeny of the chordates (1885, 3:327–30) has a main line running through to the separation of the birds and mammals, while a number of more primitive forms are treated as only slightly modified versions of what had once been major steps in the ascent of life.

Balfour's school broke up after his untimely death in a climbing accident in 1882 (Ridley 1986), but the potentially non-Darwinian implications of the link between embryology and evolution theory could not be obscured. In the work of the last British exponent of the recapitulation theory, E. W. MacBride, the embryological parallel became the basis for an explicitly anti-Darwinian progressionism (Bowler 1984b). MacBride was trained by Balfour's successor, Adam Sedgwick, and went on to become a leading exponent of Lamarckism. He stressed that evolution has a central progressive theme leading toward the higher vertebrates, and he dismissed all other types as degenerates that had failed to rise to the challenge of the environment: "It is, therefore, broadly speaking true that the Invertebrates collectively represent those branches of the Vertebrate stock which, at various times, have deserted their high vocation and fallen into lowlier habits of life" (MacBride 1914, 663). Recapitulation has here become a technique for modeling evolution itself upon the goal-directed process of individual growth — a technique that was already implicit in Haeckel's and Lankester's much earlier expositions of the theory.

As previously noted, Huxley's effective conversion to evolutionism came from his reading not of Darwin's work, but of Ernst Haeckel's *Generelle Morphologie* of 1866. The German reaction to Darwin's theory had been muted at first, perhaps because many German naturalists were already evolutionists in the sense that they accepted the gradual unfolding of a purposeful trend in the history of life. Heinrich Georg Bronn had published his own developmental view of nature (1858, trans. 1859) just before he translated the *Origin* into German. Haeckel began to take Darwin seriously in 1863 and soon adopted the theory as a component of his ostensibly materialist philosophy. Darwinism soon became an important factor in German ideological debates. Haeckel saw his materialist progressionism as a powerful weapon in the struggle for national development, and he wrote the *Generelle Morphologie* as a social as well as a biological polemic. Darwinism was soon all the rage; indeed, some later writers suggested that the theory found its true home in Germany (Radl 1930, 42; Nordenskiöld 1946, 498; for modern assessments see Montgomery 1974; Corsi and Weindling 1985).

Even more than in the British case, however, the historian encounters a problem in trying to assess the extent to which German "Darwinism" made any real use of Darwin's theory. German biologists were deeply involved with the newly developed cell theory and required that evolutionism be made compatible with this view of organic structure. Even more important was the morphological tradition, now emancipated from the overt mysticism of *Naturphilo-*

sophie yet still based on a deep faith in the orderliness of natural relationships. To a large extent, German Darwinism found its expression in evolutionary morphology, and that movement represented but a slight variation on the program already underway in the pre-Darwinian years. William Coleman (1976) showed how easily Haeckel's friend Carl Gegenbaur translated the old search for archetypal patterns into the reconstruction of evolutionary genealogies. Haeckel became the best-known exponent of this movement, and he, at least, was stimulated to move in this direction by Darwin; the end result, however, was not very Darwinian. As E. S. Russell put it (1916, 247–48), the *Generelle Morphologie* was representative "not so much of Darwinian as of pre-Darwinian thought. It was a medley of dogmatic materialism, idealistic morphology, and evolutionary theory." Yet this was the format in which many biologists in Germany and elsewhere were to learn their "Darwinism."

Haeckel welcomed Darwin's materialism because it allowed him to argue that the relations between species are the result of a natural process, not a mystical archetypal pattern. But his own approach blended natural selection with generous helpings of Lamarckism and the transcendental morphology of Goethe (see the subtitle of his popular *Natürliche Schöpfungsgeschichte* of 1868, translated as *The History of Creation*, 1876a). To the extent that Haeckel accepted a role for adaptation, he tended to stress "direct adaptation" brought about by the effects of habit, that is, simple Lamarckism (1876a, chap. 10). The struggle for existence was introduced largely as a way of eliminating the less well adapted races or species produced by the Lamarckian effect. In most of his work, however, Haeckel expressed scant interest in adaptation. His real concern was the elucidation of the morphological links that would allow a complete reconstruction of the course taken in the progressive development of life. It was Haeckel who coined the modern term *phylogeny* to denote the evolutionary history of a group. His chief evidence was not the fossil record, but the study of individual growth (for which he also coined the modern term *ontogeny*). Haeckel was the chief post-Darwinian exponent of the recapitulation theory, or what he called the "biogenetic law" — the belief that ontogeny in the individual recapitulates the phylogeny of its group.

Haeckel's work was chiefly morphological in character; recapitulation linked the patterns of development seen in the form of the growing embryo and in the past history of its type. Like Darwin, though, Haeckel shared an interest in the physiological mechanisms of inheritance and growth. The theory of heredity proposed in his *Perigenesis der Plastidule* (1876b) was supposed to be mechanistic, but in fact it had profoundly idealist overtones (Robinson 1979). Haeckel was influenced by cell theory and by Rudolph Virchow's idea that the organism is composed of cooperating cells, just as the state is composed of individual humans. Heredity was linked by Haeckel to the cell nucleus, and he believed that the molecules of protoplasm in the nucleus (the plastidules) are the real key to the question. Inheritance occurs because wave motions characteristic of the molecule are transmitted to future generations, and here Haeckel alluded to Goethe's views on the affinities of chemical elements. For all his alleged materialism, Haeckel thought that every atom has a "soul" and that the plastidule molecules have the capacity of memory. They can "remember" what has happened to them and transmit this information to influence the growth of future generations. This, of course, allowed for a Lamarckian effect in which acquired characters can be inherited. The analogy between memory and heredity was, in fact, to become a leading feature of neo-Lamarckian concepts of inheritance.

It was no accident that Haeckel, the architect of the recapitulation theory, was a Lamarckian. As Gould (1977b) has shown, recapitulation became essentially a Lamarckian rather than a Darwinian concept. If evolution works by summing up random variation, there is no reason why a modern individual should grow through stages resembling the adult forms of its ancestors. Lamarckism can justify a belief in recapitulation because it assumes that variation is an *addition* to the growth process. If changes in the adult organism (acquired characters) are to be inherited, individual growth must be speeded up so that the old adult form becomes merely a stage through which the organism must pass to reach its new state of maturity. Evolution takes place by the continued addition of new stages to growth as the old adult forms are compressed back into an earlier phase of development. Thus the individual must pass through the old adult forms as it grows, thereby recapitulating the evolutionary history of its type. Earlier exponents of recapitulation such as Chambers had explained the addition of new stages to growth as the unfolding of a divine plan. After Darwin, such explicit appeals to teleology were no longer fashionable, and Lamarckism was useful precisely because it allowed the addition of stages to growth to be seen as a purely natural process.

The recapitulation theory thus illustrates the non-Darwinian character of Haeckel's evolutionism. It does so not merely through the link with Lamarckism — after all, Darwin himself accepted the inheritance of acquired characters — but through its use as a vehicle for promoting the belief that individual growth offers a model through which evolution can be understood. For Chambers, the preordained pattern of growth controls the direction of evolution. For Haeckel, the past course of evolution dictates how the modern embryo must grow. In principle, the direction of causality is reversed so that teleology is replaced by naturalism. Yet in practice Haeckel continued to use the analogy with growth as a means of conveying the assumption that evolution must necessarily progress toward its final goal. Growth follows the pattern laid down by evolution, and since growth is progressive and goal-directed, there is an implication that evolution must share these characteristics. The most powerfully non-Darwinian (and eventually anti-Darwinian) view of evolution arose from the belief that many aspects of the history of life are governed not by haphazard geographical factors but by trends driven on toward a predetermined goal whatever the environmental changes to which the organisms are subjected. Haeckel's evolutionary morphology was essentially progressionist and was even based to some extent on a revival of the old linear image of development ascending a single hierarchy of stages toward its inevitable goal.

How can Haeckel have promoted a linear image of evolution? He is often regarded as an early proponent of the view that evolution must be represented not as a ladder but as a branching tree. Yet he was remarkably skillful in concealing the divergent character of the tree's branches when it came to depicting evolution as a purposeful process. Haeckel knew that in theory the branches follow an independent course once they have separated, but he minimized the significance of this by presenting one branch as the main stem, or trunk, of the tree and dismissing all the others as side branches. By using the growth of the human embryo as the model for vertebrate evolution, Haeckel created the impression that the "lower" modern forms correspond to a linear hierarchy through which evolution has mounted toward mankind. Russell (1916, 253) quotes part of the following passage to illustrate this tendency for Haeckel to slip back into a linear model of development similar to that used by Agassiz and Chambers:

> As so high and complicated an organism as that of man, or the organism of any other mammal, rises upwards from a simple cellular state, and as it progresses in its differentiation and perfecting, it passes through the same series of transformations which its animal progenitors have passed through, during immense spaces of time, inconceivable ages ago. . . . Certain very early and low stages in the development of man, and the other vertebrate animals in general, correspond completely in many points of structure with conditions which last for life in the lower fishes. The next phase which follows

Figure 5. "Single or Monophyletic Pedigree of the . . . Back-boned Animals," from Haeckel (1879a, vol. 2, facing p. 222).

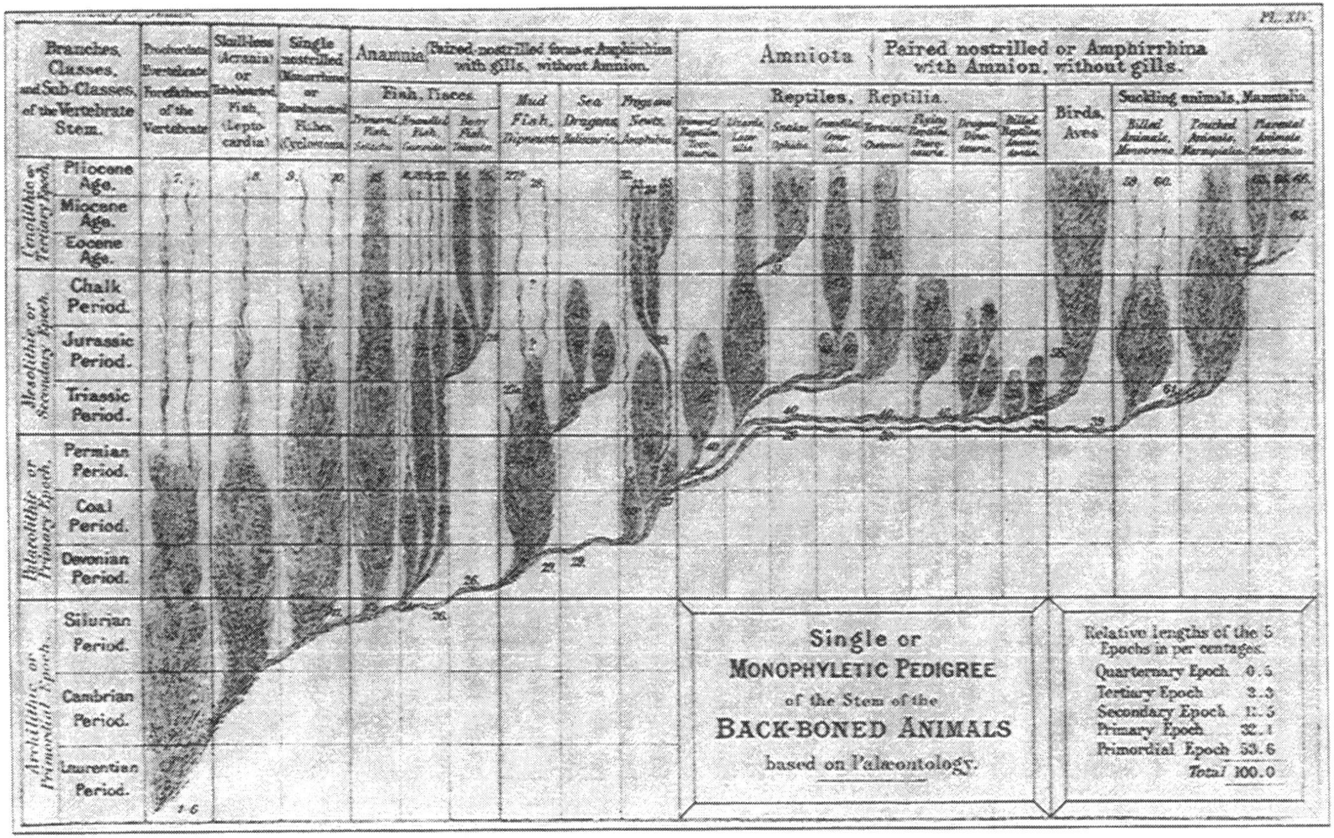

upon this presents us with a change of the fish-like being into a kind of amphibious animal. At a later period the mammal, with its special characteristics, develops out of the amphibian, as we can clearly see, in the successive stages of its later development, a series of steps of progressive transformation which evidently corresponds with the differences of different mammalian orders and families. Now it is precisely in the same succession that we also see the ancestors of man, and of the higher mammals, appear one after the other in the earth's history; first fishes, then amphibians, later the lower, and at last the higher mammals. Here, therefore, the embryonic development of the individual is completely parallel to the palaeontological development of the whole tribe to which it belongs. (Haeckel 1876a, 1:310–11)

Haeckel's most popular work, the *Anthropogenie* (trans. *The Evolution of Man*), consistently used human growth as the model for vertebrate evolution. Haeckel admitted that in its later stages the phylum divided into two branches, one leading to the reptiles and birds, the other to the mammals and eventually to mankind, but he casually asserted that it is the latter that "will in future receive our whole attention" (1879, 2:178). All too often, he went out of his way to present as many living forms as possible as relics of earlier stages in human evolution. They were the end products of side branches that did not diverge from the main stem but merely ran alongside it, preserving an ancestral form unchanged into later eras of the earth's history. The lower forms of life had merely stepped off the escalator of progress, to continue as "living fossils" that revealed earlier stages in the development of those who continued to progress. Arthur Keith said of Haeckel's interpretation of the relationship between apes and humans that the apes were seen as merely "abortive attempts at man-production" (Keith 1934, 10). The apes had *tried* to become human, but had remained stuck at an earlier stage in the process.

The clearest indication of the essentially linear character of Haeckel's evolutionism is found in the diagrams he used to illustrate the course of development, such as figure 5, taken from his *History of Creation*. Although treelike in some respects, this diagram is very different in structure from the one used by Darwin in the *Origin*. The main stem runs diagonally across toward mankind, with the vertical side branches preserving earlier stages into the present. Simplified, it gives the following view of living relationships.

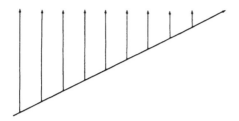

There is no real divergence, no branching in the true sense of the word, only a linear progression to the human form. A diagram used in *Evolution of Man* (see fig. 6) is deliberately drawn to resemble a gnarled tree, but it preserves the idea of a central trunk running through the whole process toward mankind, with side branches drawn in trivial proportions. In their different ways, both diagrams support the essentially non-Darwinian view that there is a central theme running through the whole of evolution and that human beings are its end product.

It is worth remembering that "evolution" was originally used to denote the growth of the embryo, and that Darwin himself seldom referred to his theory by that name (Bowler 1975). The increased use of "evolution" to denote the theory of transmutation followed the adoption of the name by progressionist writers such as Herbert Spencer and its use in the translations of Haeckel's works, such as *Evolution of Man*. There is thus a sense in which the very name of the evolution theory represents a violation of Darwinian principles, reflecting an effort to promote individual growth as a model for the development of life on the earth. The popularity of the term indicates the popularity of the progressionist and almost teleological interpretation of living development that Haeckel was promoting.

For all Haeckel's self-proclaimed materialism, his system contained a strong element of an almost mystical belief in a purposeful nature. That such an explicit progressionism was able to flourish under the name of Darwinism illustrates the extent to which the spirit of Darwin's thinking had been perverted by those who regarded themselves as his followers. His theory may have been the external influence that freed the logjam in nineteenth-century developmental thinking, but it did not control the subsequent flow of the morphological tradition. Indeed, Darwin himself seems to have been unwilling to take a stand against what Haeckel was doing. He was suspicious of the boldness with which Haeckel expressed his opinions, but he praised the chapters on the genealogy of animals in the *Generelle Morphologie* as "admirable and full of original thought"

Figure 6. "Pedigree of Man," from Haeckel (1879, vol. 2, facing p. 188).

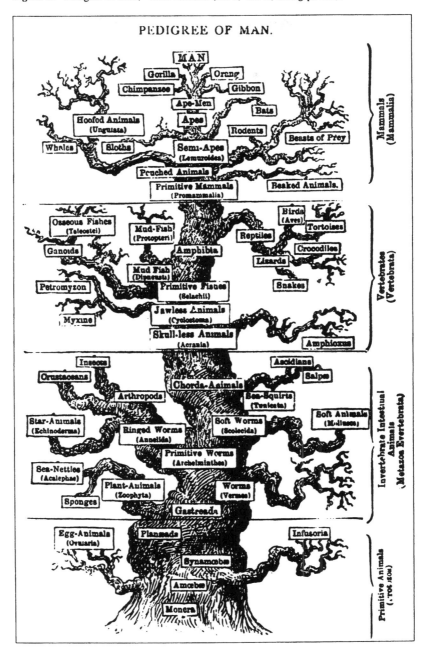

(Darwin 1887, 3:104). Darwin does not seem to have realized the extent to which Haeckel's appeal to the analogy with individual growth had imposed an almost teleological progressionism upon the theory. One can only speculate that Darwin's own continued interest in the problem of "generation" (growth and reproduction) had rendered him unable to see how Haeckel's use of this issue differed from his own. Darwin did not adopt the recapitulation theory, but he was unable to oppose it because he, too, felt that growth was important for understanding the processes of variation and heredity. In this respect Darwin remained a product of his own time; he challenged many assumptions of nineteenth-century biology, but he retained the traditional view of heredity and was thus unable to identify the source of the opposing values that so rapidly subverted the principles upon which the rest of his theory was based.

Haeckel was a Darwinian because he accepted the theory of common descent: related forms are similar because they retain the characters of their last common ancestor. His appeal to the analogy with growth led him to minimize the extent of the divergence in the tree of life, but the element of directed evolution in his thinking was expressed on so broad a scale that it was lost in the general progressionism to which almost all his contemporaries gave their support. But if growth was to be taken seriously as a model for evolution, it would have to be applied at a more detailed level, within the development of particular branches of the tree. Darwin himself had explored the view that each species has a fixed cycle of birth and death, but he had soon rejected it as he moved toward the theory of natural selection based on random individual variation. In the hands of biologists who were actively opposed to Darwinian materialism, the analogy between the life cycle of the individual and the history of its group became the vehicle for promoting a theory in which evolution is based not on divergence but on parallelism, with whole groups of species progressing together through the same pattern of development. The difference between pseudo-Darwinism and anti-Darwinism lay only in the method of applying the analogy with growth. In the later decades of the century, an increasing number of biologists came to realize that the anti-Darwinian mode of application gave a clearer expression of their commitment to the developmental viewpoint.

Anti-Darwinism

Most accounts of the Darwinian Revolution devote a good deal of space to the arguments that were raised against natural selection. Even when orthodox accounts concede that many scientists refused to accept the selection theory, however, they still manage to depict the situation in such a way that Darwin's theory remains in the spotlight. Readers are often told more about how Darwin responded to objections than about the objections themselves. Particular emphasis is laid upon those arguments that in retrospect can be seen to have pinpointed the weaknesses in Darwin's thinking which were to be filled in by later research. This, of course, paves the way for a treatment of later episodes which concentrates on the areas needed to complete the Darwinian vision. The most obvious illustration of this tendency is the fascination with the alleged deficiencies of Darwin's concepts of heredity and variation. Eiseley's (1958) account is the classic exposition of the view that Darwin's "failure" to anticipate the laws of Mendelian inheritance left his theory with a weakness that had to be overcome by later biologists. Such an interpretation misses the real point. It insists that Darwin's theory be judged by modern standards, by portraying genetics as the missing piece of the jigsaw puzzle that would have to be inserted before the picture was complete. But it is meaningless to claim that Darwin "failed" to discover Mendel's laws, as though they formed a self-evident gap in his theory. In fact, the next chapter will demonstrate that the selection theory was conceptually sound even on the basis of blending heredity. Darwin's acceptance of the traditional view of generation should be of interest, not because it prevented him from discovering genetics but because it may help to explain why the more radical aspects of his thinking were so easily subverted. The emergence of modern genetics will have to be seen as a crucial episode in the history of evolutionism, not because it filled in a gap in Darwin's theory but because it finally destroyed the older view of development that Darwin had challenged but failed to overcome.

It is thus my contention that the classic preoccupation with Fleeming Jenkin's (1867) review of the *Origin* is misplaced. According to Eiseley's interpretation, Jenkin showed that natural selection would not work within a system of blending inheritance, thus forcing the Darwinians to mark time until Mendelism emerged to solve the problem. But instead of agonizing over the defects of Darwin's theory, historians should be asking themselves what Jenkin and all the other critics wished to put in its place. Simply concentrating on the opposition to natural selection has allowed historians to ignore the fact that the opponents had very positive intentions of their own. They wished to destroy natural selection in order to erect an alternative version of how evolution works — a version that would preserve some of the values challenged by Darwin's theory. Where the pseudo-Darwinians modified the theory to retain traditional ideas, Darwin's opponents actively sought to create a vision of evolution that would repudiate his approach completely. Instead of concentrating on points that were merely intended to cripple natural selection, a historian should look to those arguments that helped to specify the kind of alternative the opponents preferred. Such an alternative was eventually created, although it took some time for the details to be worked out. The delay explains why pseudo-Darwinism gained a valuable breathing space — and why historians have failed to notice that many anti-Darwinian arguments were tending toward a particular alternative vision. Once the comprehensive anti-Darwinian theory was available, however, it gained in influence until it eventually threatened the very existence of Darwinism.

There is a very real sense in which the anti-Darwinian theories of the late nineteenth century represent the culmination of the "non-Darwinian revolution." Once roused by the Darwinian gadfly, the exponents of the morphological tradition and the developmental viewpoint set about refurbishing their ideas in a manner that would accommodate the basic idea of evolution but preserve their belief that natural development must be a purposeful and orderly process. Their first efforts were crudely expressed, since they retained an explicit role for teleology in the form of a divine superintendence of the evolutionary process. This was the "theistic evolutionism" popular among conservative scientists in the 1860s and 1870s. But explicit use of the argument from design was no longer fashionable — to this extent the Darwinians' effort to separate science and religion had succeeded — and the exponents of the developmental viewpoint were forced to seek naturalistic explanations of how evolution could be constrained to unfold in an orderly manner. They found these explanations in the theories of Lamarckism and orthogenesis, both of which used the recapitulation theory and the analogy with growth to preserve the idea that the development of life follows a determined pattern. Once these alternatives were established, the stage was set for what Julian Huxley later called the "eclipse of Darwinism" which occurred at the end of the nineteenth century (1942, 22–28).

The orthodox historiography of the Darwinian Revolution passes over the anti-Darwinian theories as irrelevant. They do not lie on the main line of conceptual development from Darwin to the modern genetical theory of natural selection, so they are dismissed as a side branch, not worthy of the historian's attention. Ernst Mayr's classic exposition of the orthodox viewpoint admits that the majority of biologists did not accept natural selection until the 1930s, yet he devotes only half a dozen pages to the alternatives (1982, 514, 525–

31). My own book *The Eclipse of Darwinism* (1983) was an attempt to rehabilitate these forgotten anti-Darwinians. That it took so long for even an outline survey of this episode to become available is a striking commentary on the imbalance created by the Darwin industry in the history of science. I now urge even more strongly that a thorough study of these theories is essential if there is to be an accurate understanding of how evolutionism developed in the nineteenth and even the early twentieth century. In some areas of biology, especially paleontology and paleoanthropology, anti-Darwinian ideas remained dominant until the emergence of the Modern Synthesis in the 1940s. They established the structure within which most efforts to understand the development of life revealed by the fossil record were conducted and hence played an important role in shaping the public perception of evolutionism. Equally significant, though, are the parallels that can be discerned between the non-Darwinian model of evolution in biology and the progressionist theories so prevalent in the nineteenth-century social sciences. Truly, if we are to understand what evolutionism meant to the nineteenth-century mind, we must abandon the view that Darwin's theory represents the pivot around which everything else moved.

What, then, were the objections to natural selection that helped to shape the anti-Darwinian movement? At least one important argument came from outside biology — Lord Kelvin's attack on Lyell's uniformitarian geology, based on the claim that the earth could not possibly have the vast antiquity Lyell attributed to it (Burchfield 1974, 1975). Using the laws of physics as then understood, Kelvin argued that the earth must be gradually cooling down and estimated its total age as one hundred million years. It is clear that Darwin's theory was one of Kelvin's real targets, since he knew that Darwin had envisaged natural selection as an immensely slow process that would need Lyellian quantities of time to achieve the results we observe. Kelvin believed that evolution must be a more purposeful process that would reach its goal more quickly. A generation of biologists were encouraged to search for non-Darwinian mechanisms by the belief that they would be necessary to make the general idea of evolution plausible within Kelvin's time scale. Only in the twentieth century did it become apparent that the discovery of radioactivity had revealed a source of heat that would keep the interior of the earth hot for a period far beyond anything that Kelvin allowed.

Kelvin was not the only physical scientist to find Darwin's theory objectionable. The astronomer Sir J. F. W. Herschel called natural selection the "law of higgledy-piggledy" (see the comment in Darwin

1887, 2:241). Herschel expressed a preference for a theory in which development was guided along regular channels, not left to the hazards of migration and adaptation.

> An intelligence, guided by a purpose, must be continually in action to bias the direction of the steps of change — to regulate their amount — to limit their divergence — and to continue them in a definite course. . . . On the other hand, we do not mean to deny that such intelligence may act according to law (that is to say on a preconceived and definite plan). (Herschel 1861, 12)

As a leading philosopher of science, Herschel was associated with the assault on Darwin's methodology (Hull 1973). It was widely argued that Darwin had failed to justify his confidence in a Lyellian restriction to processes observable in everyday life. The scientist was free to use his intuition in dealing with complex problems such as the origin and development of life, and many felt that their intuition told them that a deeper purpose must be involved.

It was up to the biologists to work out the details of a theory that would provide a worthwhile scientific expression of this intuitive feeling. Morphologists were particularly inclined to believe that the patterns they could construct to link the known living and fossil forms must somehow express a plan built into the development of life. Natural selection might play a purely negative role in eliminating the unfit, but the real problem of evolution centered on the origin of new characters. Variation was obviously the ultimate source of change, and few were prepared to believe that the everyday random variation studied by Darwin could be the source of purposeful evolution. Some more positive creative force must step in, at least from time to time, to direct the production of new characters along orderly and purposeful lines.

Many of the arguments used by biologists to attack the selection theory were intended to show that evolution must be a more actively directed process than Darwin imagined (Vorzimmer 1970). Perhaps the most effective barrage of these arguments was contained in St. George Jackson Mivart's *Genesis of Species* (1871; see Gruber 1960). This book is a veritable bible of anti-Darwinism; its objections have been repeated (if not always acknowledged) by later generations from the early geneticists to the modern creationists. But it would be misleading to concentrate on the negative aspects of Mivart's attack. His arguments were all directed to one purpose, namely, demonstrating that a purely naturalistic theory of adaptive evolution such as Darwin's could not account for the observed state of the living world or the fossil record. Mivart pointed to the existence of characters that seemed to have no adaptive purpose as well as to apparently similar characters that had appeared independently in entirely different lines of evolution. He argued that even if a structure has a useful purpose when fully formed, its early, or incipient, stages of development can hardly confer any survival value. A wing is useful for flying, but a half-formed wing is of no use for any means of locomotion. Mivart's intention was to establish the case for a non-Darwinian process of evolution that occurred not in response to the environment but through the action of forces internal to the organisms, predisposing them to vary in particular directions. He was prepared to admit that a stimulus from the environment might be needed to trigger the changes, however, since this allowed him to claim that the transmutation events must be sudden, or saltatory, thereby preserving the reality of species as distinct steps in the ladder of a group's evolution (1871, 255–56).

Mivart was able to quote with approval Richard Owen's theory of "derivation." Thanks to a vitriolic review of the *Origin* (1860), Owen has often been cited as an outright opponent of evolutionism, but the theory advanced in the third volume of his *Anatomy of the Vertebrates* made it clear that he was now prepared to accept transmutation under the guidance of a divine plan.

> Derivation holds that every species changes in time, by virtue of inherent tendencies thereto. "Natural Selection" holds that no such change can take place without the influence of altered external circumstances. "Derivation" sees among the effects of the innate tendency to change irrespective of altered circumstances, a manifestation of creative power in the variety and beauty of the results. (Owen 1866–68, 3:808).

The idea that the beauty of nature indicates an underlying creative plan was supported by other theistic evolutionists, including the duke of Argyll (1867) and William B. Carpenter (1888). Here again is the idealist version of the argument from design — the belief that the natural order is an expression of the Creator's rational will — now adapted to the possibility that the pattern might unfold gradually through a process of transmutation. Mivart was willing to admit that the design might somehow be programmed into the nature of matter itself, waiting to unfold as the right stimulus was presented, but he felt that this possibility did not diminish the value of the belief that the pattern of development is an expression of the divine will (1871, 243–45, 271). Later exponents of directed evolution tended to stress the guiding power of material forces rather than the divine will, but there can be little doubt that this view of natural development originated in the idealist philosophy of the mid-nineteenth century.

At first sight, the claim that nature unfolds according to a regular pattern seems almost vacuous, typical of a vague supernaturalism that can be linked to any factual observation. But Mivart and Owen were active in the attempt to uncover the evolutionary processes hidden by the imperfection of the fossil record. In particular, they were able to use the concept of parallel evolution to explain how several lines of reptilian development might have independently given rise to different groups of mammals (Mivart 1887–88; see Desmond 1982). Parallelism is the most important practical manifestation of the belief that evolution unfolds in a predetermined direction. In Darwin's theory, all the members of a related group owe their underlying similarity to descent from a single common ancestor. Once two species separate, there is nothing that can compel them to evolve in the same direction. But if evolution is directed by internal forces, several forms might independently move in the same direction, because they share the same internal potential. Similarities may be due to parallel evolution, not common descent. Soon a whole generation of paleontologists would be committed to the idea of parallel evolution and to the essentially non-Darwinian view that development proceeds without reference to the demands of the environment.

Mivart's vision of parallel lines advancing under the control of pre-programmed forces is a direct continuation of the style of reasoning popularized in the pre-Darwinian era by Chambers's *Vestiges of Creation*. The only factor missing is the claim that the internal force is under the control of the process of individual growth, and Mivart accepted this in a later article (1884). The chief advance that separates the post-Darwinian from the pre-Darwinian expressions of this approach is that Mivart and the later advocates of directed evolution had accepted that there is no single line of progress running through the whole history of life. Each group would have to be treated as a distinct branch of evolution under the control of its own internal forces. But was it possible to express the idea in a more scientifically respectable fashion, abandoning the vague notion of a divine plan in favor of some more naturalistic explanation of how development is controlled? This is where the analogy with growth proved particularly valuable. Growth is obviously natural (even if the mechanisms involved are not understood), and if the growth-process can somehow be seen as the agent controlling the direction of variation, the result will be an apparently natural explanation of directed evolution. It was by exploiting this possibility that the theistic evolutionism of Owen and Mivart was transformed into the superficially more scientific forms of the growth analogy that were to become popular in the late nineteenth century: Lamarckism and orthogenesis.

Two lines of thought sprang from Mivart's polemic, one far more hostile to the logic of Darwinism than the other. It was the *Genesis of Species* that inspired Samuel Butler to turn against Darwin's theory and begin the construction of his own alternative view of evolution —a view that has gained wide publicity through Butler's fame as a novelist and through the adoption of similar ideas by other literary figures such as Bernard Shaw (Willey 1960; Bowler 1983, chap. 4). Mivart had hinted that some of his criticisms might not be valid against the much stronger element of Lamarckism in Herbert Spencer's theory of evolution. Lamarckism is, after all, a process of directed variation: individuals vary not at random but by acquiring new characters (or inheriting characters acquired by their parents). Butler took up this point and realized that the inheritance of acquired characters gives an alternative way of visualizing evolution as a purposeful process. Natural selection works by trial and error, reducing the individual organism to a puppet that has no control over its inheritance or its chances of survival. Butler wanted to believe that the activity of the organism is *not* irrelevant, that its efforts to cope with problems in the environment somehow shape the course of its species' evolution. Lamarckism would allow this, since according to Lamarck acquired characters are developed in direct response to the organism's needs, as expressed in its behavior. Faced with a challenge from the environment, the organism's own initiative will lead it to adopt a purposeful new habit. If the resulting acquired characters are inherited, evolution will be directed along a channel determined by the organism's behavior. Although there would be no divine plan predetermining every step, evolution would still be a purposeful process because it would be under the control of the organisms themselves. If the spark of life is seen as a divine gift, the process can still be treated as a manifestation of the Creator's will. God, in effect, has given living things the power to design themselves. This was the view that Butler developed in books such as his *Evolution, Old and New* and *Luck or Cunning?* Since he also tried to argue that Buffon, Lamarck, and Erasmus Darwin were the true founders of evolutionism, he earned the enmity of the Darwinian community.

Butler's hatred of the mechanistic aspect of Darwinism is, perhaps, understandable. Darwin had come to believe that new behavior-patterns often arise not from the initiative of individual organisms but from selection applied to the random variation of existing instincts (Burkhardt 1985). This, of course, enhances the impression that the individual has no control over its own destiny. But it is by no means clear that Butler's position was as fundamentally anti-Darwinian as he thought. He still accepted that the organisms' efforts to adapt to changes in their environment comprise the driving force of evolution. *Any* theory that makes adaptation the determinant of change is much closer to the spirit of Darwinism than it is to Chambers's or Mivart's conception of internal forces driving species along a predetermined path. Darwin himself did not deny a limited role for the inheritance of acquired characters, and he was thus able to admit that the learning of new habits by the animals themselves can play a role.

The clearest illustration of the fact that a concern for "initiative" in evolution could be incorporated into Darwinism can be found in the story of the Baldwin effect. In the 1890s, the psychologists James Mark Baldwin and Conway Lloyd Morgan set out to incorporate Butler's point into a mechanism of evolution that did not require Lamarckism (see Baldwin 1902). They argued that acquired characters can direct evolution *even if they are not inherited.* Such nonheritable characters allow the animals to adapt temporarily to a new behavior pattern while natural selection begins to favor those individuals born with equivalent genetical adaptations due to random variation. Selection will thus follow the lead given by the new habit. The Baldwin effect blurred the distinction between natural selection and Lamarckism in a way that allowed Darwinism to be reconciled with the prevailing view that initiative and personal effort are the driving forces of evolution. Although in one sense a challenge to the trial-and-error principle at the heart of Darwin's theory, such a compromise was still incompatible with the developmentalists' emphasis on directed or predetermined evolution. It would still allow the hazards of migration and adaptation to break up the tree of life into a multitude of divergent and unpredictable branches. A number of field naturalists adopted what I have called "environmental Lamarckism" — a simple appeal to the inheritance of acquired characters as a replacement for natural selection in adaptive evolution (Bowler 1983, chap. 5, 6). In general they were well aware of the role played by geographical factors in adaptation and speciation. Although rejecting natural selection, they were thus among the relatively small proportion of biologists who were able to appreciate the more general principles underlying Darwin's challenge to the developmentalist tradition.

For this reason, there were some biologists who tried to minimize the friction between Lamarckism and Darwinism, although their efforts were overshadowed by the outbreak of hostility that brought about the eclipse of Darwinism in the 1890s. Yet this conflict was precipitated by the small number of biologists who wished to purify Darwin's vision by treating natural selection as the sole mechanism of evolution. August Weismann was the leader of these "neo-Darwinists" (see next chap.). When Weismann declared that the inheritance of acquired characters is a theoretical impossibility, his dogmatism alienated many biologists who had been prepared to go along with the more flexible version of evolutionism promoted by Darwin and his early followers. Even Herbert Spencer was forced to speak out against Weismann in defense of the inheritance of acquired characters (1887, 1893). Spencer had always accepted a major role for Lamarckism within his progressionist system, but he had acknowledged the importance of natural selection, too, and had got along quite well with the original Darwinians. Now he was forced to oppose neo-Darwinism in order to sustain the Lamarckian component of his system. Far from defending Darwinism, Weismann so polarized opinions within the biological community that the Lamarckian backlash temporarily eclipsed his own theory. Samuel Butler, who had long been ostracized by the Darwinians, was now able to report with some satisfaction that their hold on British science had been broken (see his 1890 essay "The Deadlock in Darwinism," repr. in Butler 1908).

Long before this episode, however, a very different application of Lamarckism had already begun to emerge as the foundation of a more damaging alternative to Darwinism. The emergence of this alternative can be charted most clearly in America, where the native school of neo-Lamarckism arose quickly enough to block even a pseudo-Darwinian takeover of the scientific community (Pfeifer 1965, 1974; Bowler 1983, chap. 6). Louis Agassiz, who had provided some of the most explicit pre-Darwinian expositions of the idealist argument from design, was now a leading figure in American science (Lurie 1960). He remained bitterly opposed to transmutation, but his younger disciples found it impossible to resist during the 1860s. The paleontologists Edward Drinker Cope and Alpheus Hyatt were prominent in the move to create an interpretation of evolution that would preserve Agassiz's view that nature is designed around an underlying pattern. To begin with, they explored the concept of directed evolution without offering any naturalistic explanation of how the process was guided. In an early paper, Cope openly accepted that the pattern of evolution was "conceived by the Creator according to a plan of His own, according to His pleasure" (1868, 269; repr. in Cope 1887a).

Cope and Hyatt began from Agassiz's view that the growth of the individual offers a model for the history of life on the earth. They accepted that the pattern of development revealed by a group's fossil record is recapitulated in the growth of the modern embryo. Evolution proceeds step by step through the addition of stages to individual growth, and the pattern of development is essentially predetermined and regular. Several lines of development (corresponding to distinct species) may advance through the same pattern of stages at different rates, thereby giving parallel evolution (Gould 1977b). Cope argued that the similarities between species in the same genus are a product of parallel evolution, not common descent. There is a striking resemblance between these early ideas of the American school and those advanced by Chambers and Mivart.

The Americans, however, were not content with the rather unscientific notion that the plan of development was supernaturally predetermined. In the course of the 1870s they began to argue that the inheritance of acquired characters would offer a naturalistic explanation of how the stages were added to growth and of the regularity exhibited by the evolution of many groups. Once the ancestors of a group had adopted a particular habit of life, their bodily structure would adapt to the new behavior-pattern. The resulting acquired characters, if inherited, would guide the group's evolution toward an even higher degree of specialization for the chosen life style. All species within the group would advance through the same pattern of specialization in parallel. Cope, in particular, was alive to the theological implications of this position, and he argued more or less along the same lines as Butler: Lamarckism vindicates the belief in a divine purpose expressed through the activity of living organisms (Cope 1887b). The Creator has delegated to the organisms themselves the power to direct their own evolution. The progress of life is inevitable, since increased intelligence always gives a better response to the environment and will thus be favored by the Lamarckian process. The key difference between Butler and Cope is that the latter saw the initiative of the founders of a group as a determining factor that will direct the whole of the group's subsequent evolution through a linear pattern of specialization.

The American biologists were not engaged in idle speculation. Cope and Hyatt were industrious paleontologists; Cope, indeed, is best known for his feud with O. C. Marsh over the opening up of the fossil beds of the American West (Plate 1964; Lanham 1973; Shor 1974). Cope saw the apparently linear pattern of specialization displayed by many vertebrate families as evidence for the non-Darwinian character of evolution. Although Marsh and Huxley presented the fossil evidence for the evolution of the horse family as favorable to Darwinism, Cope argued that the fossils revealed a trend too regular to be explained by any mechanism based on random variation. Cope also believed that many evolutionary trends could not be explained by adaptation or specialization. They produced nonadaptive characters and would have to be explained as the result of some purely internal driving force.

Hyatt (1866, 1884, 1889) made even more extensive use of the concept of nonadaptive evolution. Working with fossil cephalopods such as the ammonites, he extended the analogy between evolution and growth to include the senility and death of whole invertebrate groups. Since ammonites preserve the immature stages of growth in the internal segments of their coiled shell, Hyatt was able to show that growth did indeed seem to recapitulate evolutionary development. But he also saw a complete life cycle in the history of the group as well as in the growth of the individual. A single organism develops progressively up to the point of full maturity but then runs out of growth-energy and declines toward old age and death. And so, Hyatt argued, a whole group of related species can undergo a preliminary phase of progressive evolution under the influence of the Lamarckian effect, but the group will eventually use up all of its evolutionary energy. After that, all members of the group will enter a phase of degenerative evolution, declining in parallel through stages of increasing simplicity toward racial senility and extinction. The pattern of apparent degeneration revealed by Hyatt's cephalopods continued to puzzle biologists well into the twentieth century and was widely taken as a classic illustration of non-Darwinian evolution.

Hyatt's concept of racial senility exerted little influence during the heady days of pseudo-Darwinian progressionism, but by the 1890s biologists were beginning to take nonadaptive evolution more seriously. The German biologist Theodor Eimer popularized the term *orthogenesis* to denote these nonadaptive trends (Bowler 1979; Bowler 1983; chap. 7). Significantly, Eimer adopted a Lamarckian approach to adaptive evolution, linking the mechanism to the process of individual growth (trans. 1890). Like many others at the time, he regarded Weismann's dogmatic neo-Darwinism as a wild speculation and redoubled his efforts to undermine the basic principles of Darwinism. Eimer's insistence that many aspects of evolution are nonadaptive was meant to back up this position (trans. 1898). The term *orthogenesis* was increasingly taken up by paleontologists who, like Hyatt, were convinced that they could see nonadaptive trends in the fossil record (Rainger 1981). The most powerful manifestation of this idea was the theory of overdevelopment, according to which the evolutionary growth of certain once-useful organs would gain a "momentum" that would eventually lead to excessive growth and finally, extinction. The massive antlers of the so-called Irish elk were often cited as an illustration of this kind of orthogenetic trend. Here again the analogy with growth was used to argue that evolution is driven toward a predetermined goal, whatever the demands of adaptation. The theory of orthogenesis is the most extreme form of anti-Darwinism. It not only replaced common descent with parallel evolution in the explanation of organic affinities, but also repudiated the whole Darwinian emphasis on the power of adaptation to shape evolution.

The burden of trying to reconstruct the history of life on the earth passed steadily from pseudo-Darwinians such as Huxley and Haeckel to anti-Darwinians such as Cope, Hyatt, and their followers. Lamarckism and orthogenesis were to remain popular among paleontologists until well into the twentieth century. The prominent American paleontologist Henry Fairfield Osborn continued to use the concept of nonadaptive trends under a variety of different names (e.g., 1936). To the extent that many nonscientists saw phylogenetic research as the most important area of evolutionism, Darwin's theory fell into the background. Even the origins of the human race were interpreted largely in non-Darwinian terms into the 1930s (Bowler 1986). The paleontologists and paleoanthropologists remained unmoved by Weismann's theory of the germ plasm and by the emergence of modern genetics. There was an almost complete breakdown of communication between the new experimental disciplines and traditional areas of biology such as paleontology. In Germany, where dogmatic Mendelism did not gain such a stranglehold over the scientific community, non-Darwinian ideas remained popular into the mid-twentieth century, culminating in O. H. Schindewolf's typostrophe theory (Reif 1983, 1985). By simply ignoring these developments in favor of Mendelism and the genetical theory of natural selection, the historians of the Darwin industry have created a completely artificial picture of how scientific evolutionism developed.

The paleontologists' fascination with linear trends can be attributed to several causes. Rainger (1985) argued that in its original form, paleontology was essentially a morphological discipline. It was based on a study of the structure of fossil specimens, and this naturally encouraged the view that its chief theoretical objective should be the construction of hypothetical links between related forms. Since most groups were represented by very few specimens, it was only too easy to produce oversimplified patterns that would not be disproved until further discoveries revealed the true (i.e., irregular and hence Darwinian) character of the group's evolution. The supposedly non-Darwinian trends were simply the product of a failure to appreciate the imperfection of the record, not of any overwhelming faith in non-Darwinian mechanisms. Later exponents of parallel evolution and orthogenesis certainly tended to avoid the strident antimaterialism of Mivart and the mid-nineteenth-century school of idealist morphology. Yet their anti-Darwinian arguments were frequently little more than a direct continuation of the style of reasoning which had been popular in the earlier period. It was now fashionable to argue that the properties of organic matter — not a divine plan — directed evolution along fixed channels. Mivart, too, had conceded that the Creator's plan might be built into the basic properties of the material universe, merely waiting to unfold in the course of time. The theory of orthogenesis was a direct continuation of a pre-Darwinian way of thinking, and the clearest illustration of this can be seen in the continued use of the analogy between evolution and growth. I have described how both Chambers and the American neo-Lamarckians used the recapitulation theory to argue that evolution proceeds by the preordained addition of stages to individual growth. The anti-Darwinian theories of the period around 1900 must be treated as the final product of the developmental viewpoint that had flourished throughout the preceding century.

Darwin may have precipitated the morphologists' conversion to evolutionism, but he exerted little control over how they used the idea. The developmental view of nature continued, and the growing power of anti-Darwinian at the expense of pseudo-Darwinian theories is clear evidence of the superficiality of Darwin's impact. The eclipse of Darwinism was an inevitable consequence of this superficiality. Once the shock of the Darwinian challenge had been absorbed, the developmentalists began to formulate versions of evolutionism that would preserve some aspects of their original world view. Within a morphological discipline such as paleontology, non-Darwinian ideas would naturally be more attractive, and it was only a matter of time before this non-Darwinian character became explicit. Once an anti-Darwinian rival was established, the fate of pseudo-Darwinism was sealed. When a German critic wrote of *The Deathbed of Darwinism* (Dennert 1904), he was expressing no more than the simple truth. There was no longer any need to pay even lip service to the theory of natural selection.

Yet the morphological approach to biology was itself coming under threat from a new breed of experimental researchers. In the early twentieth century these scientists pioneered Mendelian genetics, which effectively discredited the analogy between evolution and growth. The conceptual framework within which both pseudo-Darwinism and the developmental form of anti-Darwinism had flourished was now undermined. Inspired by the moral arguments in favor of their theory, some Lamarckians made a valiant effort to retain a place within the new experimental study of heredity. But direct evidence for the inheritance of acquired characters had always been elusive, and the geneticists preferred a model of "hard" heredity over which the environment had no control. Once divorced from its link with the growth analogy, simple Lamarckism proved difficult to defend. The clearest illustration of this is the "case of the midwife toad," in which the Austrian biologist Paul Kammerer (1923, 1924) tried to provide experimental support for Lamarckism. Kammerer was eventually dismissed as a fraud by the geneticists and committed suicide. Arthur Koestler's (1971) account of this affair paints a tragic picture of a scientist faced with an increasingly hostile climate of opinion. Although he failed to mention that Lamarckism had been flourishing a few decades earlier, Koestler correctly sensed that in the English-speaking world, at least, the emergence of genetics marked the end of Lamarckism's credibility.

The geneticists thus succeeded where Darwin had failed. They destroyed the analogy between growth and evolution upon which theories such as Lamarckism had depended. Only in areas such as paleontology, where the morphological tradition remained untouched by the demand for experimental evidence, did these theories linger on into the early twentieth century. Unfortunately for the few true Darwinians, the geneticists were infected with a distrust of the older sciences which led them, too, to reject Darwinian principles. They threw out the Darwinian baby along with the pseudo-Darwinian bathwater. In the 1920s the historian of biology Erik Nordenskiöld hailed the emergence of the new experimentalism, but he saw no prospect of the revival of Darwinism ([1929] 1946, 528, 161). The eclipse of Darwinism was thus compounded: the final stages of the Non-Darwinian Revolution in morphology coincided with the anti-Darwinian birth pangs of an entirely new science. What Nordenskiöld did not realize, however, was that the application of experimental techniques to the study of heredity had produced a revolution that would ultimately allow natural selection to emerge as the most plausible mechanism of evolution. Genetics, as it turned out, was intrinsically far more hostile to the growth analogy than to Darwinism. The Mendelian Revolution would sweep away the whole framework within which the non-Darwinian approach had flourished. Out of the resulting chaos would eventually come the realization that something along the lines of Darwin's original insight offered the only way of turning laboratory genetics into a workable theory of real-life evolution.

Chapter Five

FROM DARWIN TO MODERN DARWINISM

✤ If evolutionary morphology was either pseudo-Darwinian or anti-Darwinian in character, then, to the extent that it constituted the dominant form of late-nineteenth-century evolutionism, Darwin's theory enjoyed little real influence. The *Origin of Species* had precipitated the Non-Darwinian Revolution within the morphological tradition, but what is now recognized as the book's most important message was of little interest at the time. Natural selection was ignored or vilified. Even the more general principles of Darwin's biogeographical approach were not taken up by the anatomists and paleontologists who favored the developmental view of evolution. For these scientists, morphology remained influential into the early decades of the twentieth century, fending off the challenge from the brash new experimental science of genetics. The developmental view of evolution was only eliminated completely in the 1930s, when genetics was synthesized with a revived Darwinism to give an evolutionary mechanism capable of winning support in all areas of biology. Few historians of science would now dispute the claim that it was not until this decade that a majority of evolutionary biologists became, in the modern sense of the term, Darwinians.

There is thus a sense in which the emergence of the modern synthetic theory can be seen as the first real triumph of Darwinism. If there was a Darwinian Revolution, it was not completed until the 1930s, and then only through the intervention of a new theory of heredity. But should this later episode be regarded as the culminating phase of the *Darwinian* Revolution? The answer to this question depends on the extent to which the various factors making up the genetical theory of natural selection were produced directly or indirectly in response to arguments advanced in the *Origin of Species*.

In the interpretation offered by Eiseley (1958), the twentieth-century developments were necessary if the essential principles of Darwin's theory were to be clarified. Darwinism could not work without Mendelian genetics, and thus it was inevitable that this "gap" in Darwin's original formulation of his theory would have to be filled in. Genetics constitutes the missing piece in the Darwinian jigsaw puzzle. The origin of Mendelism can thus be treated as an extension of Darwinism — and the final phase in the Darwinian Revolution.

The fact that Eiseley's interpretation ignored the strength of anti-Darwinian feeling around 1900 indicates that it is a form of Whig history, a view of the past colored by modern preconceptions about how evolutionism ought to have developed. It should by now be obvious that nineteenth-century evolutionism cannot be understood solely in terms of factors that pave the way toward modern ideas. But Eiseley's view of twentieth-century Darwinism may also represent a Procrustean bed into which complex developments have been trimmed to fit by the cutting edge of hindsight. There are certainly some areas of biology in which reconsideration of issues raised by Darwin helped to shape modern evolutionism. But the most innovative force in early-twentieth-century biology — Mendelian genetics — was introduced with the aim of replacing, not reforming, the theory of natural selection. Only when the Mendelians moved out of the laboratory to create a science of *population* genetics did it gradually become apparent that Darwin's insights had not, after all, been as wide of the mark as many biologists had at first believed.

Rather than viewing the emergence of genetics as a direct consequence of the original Darwinian debate, it may be more fruitful to regard it as a second, even more fundamental, conceptual revolution, undermining attitudes that even Darwin had not challenged. The importance of Mendelism for evolution theory was as much negative as positive; it paved the way for the creation of modern Darwinism by weakening the analogy with growth which had been central to the developmental viewpoint. The old form of non-Darwinian evolutionism now became implausible, and the Mendelians had to look for new alternatives. Originally they had no intention of reviving the selection theory; indeed, they used anti-adaptationist arguments borrowed directly from Mivart and the supporters of orthogenesis. Many early geneticists would have preferred a theory in which evolution was directed by internal forces such as mutation-pressure (a tendency for mutations to occur consistently in a particular direction). Eventually the experimental evidence made it clear that mutations are not directed. Variation was random, as Darwin had supposed, and the attempt to show how genetic factors are controlled within a wild population led back to natural selection as the only viable directing agent of evolution.

An overzealous historian of genetics might argue that modern biology would have been forced to create the theory of natural selection if Darwin had not suggested it earlier. From this perspective, Darwin would appear as merely the precursor, not the true founder, of modern evolutionism. The most significant break with the past would be the Mendelians' destruction of the growth analogy, from which the selection concept would emerge almost as a casual by-product. This position is, of course, a caricature, although it is worth mentioning as an antidote to the equally misleading opinions in Eiseley's interpretation. Whatever the origins of Mendelism, its application to wild populations was made possible by statistical techniques introduced in direct response to Darwinism (Provine 1971). Ernst Mayr (1959c, 1976) insisted that modern evolutionism was also inspired by the field naturalists' exploration of a Darwinian approach to the species concept. The Modern Synthesis incorporated not just two theories (Darwinism and Mendelism) but a whole series of different research traditions. All of these traditions can be traced back to the late nineteenth century, and all were influenced to a greater or lesser extent by Darwin's work. My task is to paint a balanced picture that neither ignores nor exaggerates his role in the development of modern Darwinism.

Population Thinking

As a leading modern Darwinian, Ernst Mayr (1964, 1982) has proclaimed that one of Darwin's most fundamental innovations was the introduction of "population thinking" — the willingness to think of a species not as a fixed morphological type but as a population of diverse individuals whose average character can be changed by selection. The typological view of species is clearly part of the traditional world view, since it dovetails neatly with the assumption that each specific form was designed by God. Many non-Darwinian theories of evolution retained the belief that each species is a distinct type by assuming that each step in the pattern of development takes place by a sudden leap, or saltation. Darwin's decision to treat adaptation to an ever-changing environment as the sole cause of evolution implied, at least in principle, a focus on the breeding population as the key unit of change. Continued exploration of the link between the population and its environment thus constituted an obvious extension of the research program that had led Darwin to the most radical aspects of his theory. Such an approach did not openly challenge the analogy between growth and evolution; it simply ignored the question of individual growth to concentrate on an entirely different way of thinking about natural change.

It is significant that Darwin himself was unable to make the complete transition to a populational view of species (Beatty 1982). He still tended to think of the species as a morphological type from which occasional individuals diverged through some disturbance of their growth-process. The transition from Darwin's original theory to modern Darwinism is marked by the unpacking of the populational view of species implicit in the concept of natural selection. The claim that later biologists have "clarified" Darwin's thinking on this issue is, of course, a product of hindsight. It implies that today's interpreters know what was really important in the theory even if Darwin and his contemporaries did not. Yet the risk of writing Whig history seems less severe in this area. The small number of biologists who explored the populational view of species in the late nineteenth century were self-consciously engaging in a study of issues and conceptual approaches defined by Darwin. If they found themselves forced to differ from him, they did so gradually and without a sense of making a complete break with the past. Hindsight or no, the gradual emergence of the populational view of species represents the clearest example of Darwin's original insights leading to the establishment of research programs that are an ongoing part of modern Darwinism.

The weakness of Darwin's immediate impact on biology is revealed by the need to distinguish two different avenues through which the populational view of species was explored. The most direct continuation of Darwin's own initiative can be seen in the comparatively small number of naturalists who ignored morphology to concentrate on biogeography and the problem of speciation. In some cases a "Darwinian" view of divergent evolution was supported by naturalists who had little interest in the details of how natural selection worked. This approach did not at first interact with the biometrical school of Darwinism, which applied statistical techniques directly to the study of variation within a population. Far from initiating a unified research program among his followers, Darwin's theory offered various facets upon which a study of evolution could be built. The complete integration of these techniques would not be possible until the synthesis with genetics in the 1930s.

Darwin's experiences on the Galapagos alerted him to the evolutionary significance of populations that adapted to their local environment. Geographical distribution remained a fertile source of evidence and was perceived as one of the Origin's most original contributions to the debate. It was no accident that Darwin first revealed his ideas privately to J. D. Hooker and Asa Gray, both field naturalists with a strong interest in geography. Hooker's introduction to his study of the flora of Tasmania (1860) was one of the first pro-Darwinian scientific works (Turrill 1963; Allan 1967). Gray led the American defense of Darwinism and wrote a number of articles applying the theory to the distribution of plants in America (1876; Dupree 1959). As a student of animal distribution, A. R. Wallace published a series of works illustrating the Darwinian approach (1869, 1870, 1876, 1880), eventually becoming more "Darwinian" than Darwin himself. He defined "Wallace's line," which separates the Asian from the Australian fauna in Southeast Asia on the basis of the possibilities of migration in past eras when the sea level was lower than at present (Beddall 1969; Mayr 1954; Fichman 1977).

Strong evidence also came from the study of animal coloration. Henry W. Bates (Wallace's early traveling companion) discovered a new form of insect mimicry, in which an edible species copies the warning colors of an inedible form to gain protection from predators (Bates 1862; Woodcock 1969). This could not be explained by Lamarckism, since insects have no control over their coloration. Later in the century, animal coloration was used by Edward B. Poulton (1890, 1908) to defend natural selection at a time when it had come under increasing pressure. Field naturalists were aware of the close relationship that must exist between a population and its environment and were less likely to be seduced by theories invoking large-scale nonadaptive trends. There was, however, a widespread belief that some minor characters have no adaptive significance, a view that created problems for Darwinism well into the twentieth century (e.g., Robson and Richards, 1936).

There was another area of confusion that hindered the field naturalists' support for Darwin. Inspired by his Galapagos studies, Darwin originally believed that geographical isolation was essential for speciation. Only when physically prevented from interbreeding would related populations acquire different adaptations that might eventually turn them into separate species. By the time he wrote the Origin, though, Darwin had become convinced that ecological specialization across a continuous territory could split a population into fragments that would become separate species (Mayr 1959b; Ospovat 1981; Sulloway 1979a). He was challenged on this point by Moritz Wagner (trans. 1873), who insisted that geographical isolation was essential to prevent interbreeding that would blend potentially divergent characters back together again. Darwin would not accept Wagner's claims, but was unable to explain how the tendency to blend would be overcome. This left the way clear for anti-Darwinian biologists to propose other mechanisms, including saltation, to explain speciation without isolation. Even some of Darwin's supporters compromised by invoking new mechanisms of speciation, including G. J. Romanes's physiological selection (Lesch 1975).

Modern Darwinians argue that population thinking was implicit not just in the theory of natural selection but in Darwin's whole geographical approach to adaptive evolution. But the confusion over speciation suggests that Darwin himself was unable to complete the transition to a populational view of species. The confusion was only cleared up as a later generation of naturalists began to think more carefully about the issues that Darwin had raised. Ernst Mayr (1955, 1976) singled out Karl Jordan's work at the turn of the century as an important step toward the modern viewpoint. The importance of geographical isolation was now increasingly recognized, allowing adaptation to the local environment to be seen as the most important factor in evolution. This move by the field naturalists did not immediately lead to the singling out of natural selection as the only mechanism of adaptive evolution. Mayr notes that Jordan himself accepted a role for Lamarckism, as did many early-twentieth-century naturalists. Thus, continued geographical studies revitalized a "Darwinian" approach to evolution without any reformation of the selection theory. Only in the 1930s did field workers such as Mayr and Bernhard Rensch begin to appreciate that genetics had now tipped the balance against Lamarckism, leaving selection as the only viable mechanism of adaptation.

Accepting that the study of biogeography and speciation can be seen as the continuation of a research program initiated by Darwin, it is still necessary to ask whether the field naturalists' efforts could have brought about a more general move toward Darwinism in the early twentieth century. There were certainly a few paleontologists who shared a concern for geographical factors in the history of life, most notably W. D. Matthew (Rainger 1986). But Matthew's efforts were still overshadowed by the tradition of reconstructing phylogenies within which the theories of orthogenesis and parallelism continued to flourish. Biogeography by itself was not enough to destroy the developmental approach, except within its own discipline. The broader collapse of developmentalism was precipitated by studies of the evolutionary mechanism itself. Not surprisingly, most historians have focused on the emergence of genetics as the most crucial breakthrough in this area. The creation of the modern genetical theory of natural selection would not have been possible, however, without the statistical techniques developed by a handful of mathematicians and biologists who decided that the problem of evolution could be solved by a direct study of changing variability within large populations.

This biometrical approach was not completely isolated from bio-geography, since A. R. Wallace included studies of variation within wild populations in his *Darwinism* (1889). But in general the two areas developed independently. It was Darwin's cousin Francis Galton who pioneered the application of mathematics to the analysis of variation. Inspired in part by the selection theory, Galton became convinced that heredity played an important part in human affairs (Swinburne 1965; Froggatt and Nevin 1971; Provine 1971; Cowan 1972a, 1972b; De Marrais 1974; Forrest 1974; Robinson 1979). He saw the importance of maintaining valuable characters within the population while eliminating all that were harmful. Statistical techniques were introduced to estimate the variability of large populations and to measure the effect of selection over a number of generations. Although Galton's immediate followers were critical of genetics, the end product of biometry was a mode of analysis that would eventually be applied to the genetical structure of wild populations.

Galton took the theory of natural selection as his starting point, but he reworked its foundations in a way that challenged conventions to which even Darwin had remained loyal. Once again, the decision of a small number of scientists to concentrate on the more radical aspects of Darwin's thought led to a ruthless exposure of the conceptual limitations within which the original selection theory had been formulated. As has been noted in earlier chapters, Darwin retained the belief that "generation" is central to the evolutionary process. Selection might determine which variations survive, but such variation is essentially a modification of an individual organism's growth. Far from challenging the link between evolution and growth, Darwin's theory of pangenesis sought to explain heredity and variation in terms of traditional concepts. Galton rejected pangenesis and soon despaired of ever producing a satisfactory explanation of how variations are inherited. He turned his back on Darwin's fascination with growth and determined to study heredity indirectly by charting the transmission of variability within the whole population. He later wrote that

> it seemed most desirable to obtain data that would throw light on the average contribution of each ancestor to the total heritage of the offspring in a mixed population. This is a purely statistical question, the same answer to which would be given on more than one theoretical hypothesis of heredity whether it be Pangenetic, Mendelian or other. (Galton 1908, 308).

In its own way this move paralleled the complete break of genetics with the past. The analogy between evolution and growth was now irrelevant; evolution was solely a function of the changing proportions of characters within a population.

Consistent with his statistical approach, Galton formulated his "law of ancestral heredity." He supposed that each individual obtains diminishing proportions of its inheritance from its more remote ancestors — half from the two parents, a quarter from the four grandparents, and so on. Curiously, Galton himself did not believe that selection could permanently alter a species. He openly argued for a saltatory explanation of the origin of new characters (1889, 18–34). His followers, however, turned the biometrical school into a major source of support for Darwinism in the troubled years at the turn of the century (Norton 1973). Karl Pearson (1896, 1898, 1900) showed that the law of ancestral heredity would allow selection to be effective in permanently shifting the mean of a continuous range of variation. W. F. R. Weldon (1894–95, 1898, 1901) provided experimental evidence that selection could have a small but permanent effect on a wild population.

By turning their backs on the physiological problem of how new characters are formed, the biometricians separated evolution theory from the question of growth and undermined the teleological associations that Haeckel and others had extracted from the link. Evolution became a process that took place within populations, not in individuals — something the modern Darwinian takes for granted. Equally important was the emergence of a new insight into the relationship between heredity and variation. Darwin had remained true to the traditional view of heredity as an essentially conservative force, tending to make the offspring grow into a copy of its parents. Variation was an antagonistic factor, constantly threatening to disturb the individual's growth so that it acquired characters different from those of its parents. Biometry eliminated this sense of antagonism, presenting variation and heredity as merely two different aspects of the same process going on within the population. Variant individual characters were not abnormal; they were an integral part of the range of variation existing naturally within the population, preserved and recombined from one generation to the next by heredity. Growth was now irrelevant to evolution, since variation was a function of the population, not of how the individual developed its particular characters.

Historians of modern evolutionism acknowledge the importance of Pearson's work in providing the statistics that would be used in the creation of population genetics. But the overall significance of biometry's break with the traditional link between growth and evolution has seldom been recognized. Pearson's impact on the biology of his own time was limited, partly because his political opinions left him isolated from the British scientific community (Kevles 1985). More generally, biometry's role is dismissed as secondary because its emphasis on the variability of whole populations led its supporters to reject what is now perceived as the most important breakthrough in the science of heredity — Mendelian genetics. Galton's law of ancestral heredity was wrong, and a satisfactory alternative was only developed by studying discontinuous variation in small laboratory samples. Mendelism is recognized as the more positive contribution because it offers the "correct" model of heredity. This model would eventually be applied to whole populations, but only after its basic laws had been worked out under conditions bearing no resemblance to those that Darwin and the biometricians recognized as crucial for evolution.

The Mendelian Revolution

No one doubts that Mendelism helped to create the framework within which modern Darwinism operates, but opinions may differ on the historical relationship between the two theories. The tendency to focus on Darwin's work as the important breakthrough in modern biology automatically limits Mendelism's significance. In Eiseley's interpretation (1958), Mendel held the "key" to evolution. His laws of heredity were the missing piece in Darwin's jigsaw puzzle; the potential value of natural selection could not be realized without the correct model of heredity. Mendelism expresses a view of heredity which is really implicit in Darwin's theory, although in this area Darwin himself was unable to penetrate the veil of obscurity woven by traditional misconceptions. Pressure from evolutionism thus inevitably led to a breakthrough in the study of heredity, thereby clarifying Darwin's insight and completing the revolution he had begun. To support this interpretation, the work of August Weismann is singled out as a link between Darwin and the origins of Mendelism. It was Weismann who realized that natural selection demands a concept of "hard" heredity in which characters are transmitted from parent to offspring without being affected by changes in the parents' bodies. This paved the way for recognition of Mendel's laws explaining how the transmission takes place.

Without denying that Darwinism helped to stimulate research in heredity, it is possible to see the emergence of Mendelian genetics as an event of far greater significance. The new science was not created to solve the problems of Darwinism; indeed, Pearson's work showed that it was quite possible to create a workable model of natural selection using pre-Mendelian concepts of heredity. Although Weismann recognized the need for a theory of hard heredity, he did so within a conceptual framework that, like Darwin's, did not threaten the traditional link between growth and evolution. This link was finally severed when the rediscoverers of Mendel's laws turned the study of heredity into an experimental science. Genetics presented itself as an entirely new approach to issues that had been merely confused by the older disciplines within which Darwinism had been conceived. No one at first realized that it would eventually render natural selection more plausible. The re-emergence of Darwinism was made possible because Mendelism destroyed the developmental viewpoint that Darwin had left largely intact. From this perspective the "Mendelian Revolution" represents a break with tradition of a magnitude fully equivalent to that of the original impact of the *Origin.*

Eiseley's interpretation subordinates Mendelism to Darwinism from the start by insisting that the weakness of Darwin's views on heredity was the only significant stumbling block to his theory's general acceptance. Darwin's hypothesis of pangenesis (published in 1868 but conceived much earlier) not only allowed a role for Lamarckism but also preserved the traditional concept of blending heredity. By supposing that parental characters are simply blended together in the offspring, Darwin missed the significance of Mendel's discovery and left selectionism with a dangerous weakness. Fleeming Jenkin's review of the *Origin* (1867) pointed out that blending made natural selection unworkable; individuals born with significant new characters would not affect the species because their characters would be diluted or swamped by continued interbreeding with unchanged individuals. Only a theory of particulate inheritance could evade this problem, and because Darwin was unable to take this step he retreated even further into Lamarckism.

The most obvious fault with this argument is that blending heredity is *not* fatal to natural selection (Bowler 1974). If favored variants are rare individuals, Jenkin's argument is valid; but if they form one extreme of a continuous range of variation within the whole population, the favored character cannot be swamped. This point was emphasized by Wallace and was incorporated firmly into Pearson's biometrical theory of natural selection, which was based on Galton's refinement of the blending model. The selection theory was not conceptually unsound in the late nineteenth century. Its lack of influence was due not to a missing piece in Darwin's thinking but to his inability successfully to displace the prevailing belief that a study of embryological growth was of prime importance to evolutionism. The orthodox view of the history of Darwinism misses this point because it focuses too narrowly on Darwin's lack of a particulate theory of heredity. Evolutionism did stimulate research into heredity, but the field was also revitalized by the emergence of cell theory and, later, by the decision to study the transmission of characters in artificial breeding experiments. The research that would ultimately transform evolutionism began in an area of study to which Darwin was deeply committed, but it was completed only when that area was fragmented in a way that Darwin himself would not have understood.

The theory of pangenesis arose from Darwin's early interest in the phenomena of reproduction (chap. 2 above). His study of adaptive evolution was from the start conditioned by his belief that an understanding of growth was vital for any theory of variation and heredity. Darwin saw variation as a disturbance of the growth-process by external influences, and pangenesis allowed for both random changes and the inheritance of acquired characters. The "gemmules" that transmitted characters to the offspring were supposed to bud off from the appropriate parts of the parents' bodies before transmission to the reproductive organs. Pangenesis symbolizes Darwin's commitment to the idea that evolution is essentially an extension of individual growth. He may have escaped the assumption that growth predetermines the course of evolution, but he could never repudiate the link with embryology that made the recapitulation theory seem so plausible to his contemporaries.

The logic of Darwin's model was challenged by the biometricians' move to an emphasis on variation as a function of the population. But this merely evaded the question of a physiological mechanism for heredity and variation. The more obvious break with the tradition to which even Darwin had remained loyal came through efforts to understand how characters are transmitted from parent to offspring. Here the leading critic of the pangenetic hypothesis was August Weismann, whose contributions to the development of evolutionism have been hailed by Ernst Mayr (1985) as second only to Darwin's. The search for the roots of modern Darwinism may, however, have created an artificial image of Weismann's position. Other historians now stress the extent to which Weismann functioned within the same complex of ideas as Darwin himself (Hodge 1985b). Unlike the later geneticists, Weismann shared Darwin's interest in biogeography and the problem of speciation. More significant, though, is his continued reliance on the link between evolution and growth. Weismann's concept of the "germ plasm" (trans. 1891–92, 1893a) formalized the notion of hard heredity as it is now understood by modern Darwinians, but it originated from studies conceived within the developmental tradition.

Stephen J. Gould (1977b) has shown that Weismann's early work (trans. 1880–82) used the recapitulation theory to trace the evolutionary links among caterpillars with related color schemes. More recently Frederick Churchill (1986) argued that Weismann's belief that the germ plasm must be transmitted unchanged from one generation to the next was derived from studies of the Hydromedusa (polyps) in the period 1878–83. These investigations also used the recapitulation theory to argue that the adult body (the somatoplasm) acts merely as a host for the perpetuation of immortal germinal material. Weismann derived this view from his belief that single-celled organisms, which reproduce by simple division, are potentially immortal. More complex organisms evolved to take advantage of the division of labor allowed when cells specialize for different bodily functions. Only the germ cells retain the immortality of their single-celled ancestors. The purpose of sexual reproduction was now seen as the maintenance of variability within the species through the constant recombination of parental germ plasms. This paralleled the biometricians' recognition that variation should be treated as a function of the whole population, but Weismann was unable to anticipate the modern use of this idea because, as Mayr (1985) admitted, he did not abandon a typological view of species.

After 1883 Weismann took increasing note of advances in cell theory which suggested that the germ plasm consists not of the whole germinal cell or gamete but is concentrated in the rodlike chromosomes of the cell nucleus (Robinson 1979; Farley 1982; Olby [1966] 1985). This is often seen as an anticipation of the modern view, but Weismann's description of the germinal units was highly speculative and illustrates the gulf still separating his thinking from that of the early geneticists. Far from developing a concept equivalent to genetical mutation, he introduced a theory of "germinal selection" in which character-determinants compete for nourishment and only the strongest survive to produce visible characters in the adult organism (1896). Although intended as an extension of the selection mechanism, this theory created the possibility that nonfunctional characters might be produced by variation trends originating within the germ plasm. In effect, germinal selection allowed for small-scale orthogenesis.

These facts must be borne in mind during any attempt to evaluate the significance of Weismann's contribution. With hindsight it is easy to single out his insistence that the germ plasm is isolated from the rest of the body as a major clarification of the Darwinian viewpoint. In Weismann's theory, as in modern molecular biology, the information flow is one-way only; the germ plasm determines how the body will grow, but it cannot be affected by the body that transmits it. This certainly led Weismann to repudiate Lamarckism and proclaim the "all sufficiency of natural selection" (1893b). He performed a classic experiment, cutting off the tails of generations of mice to show that there was no tendency for the "acquired character" of taillessness to be inherited. Along with A. R. Wallace, Weismann thus became a founder of what was known as neo-Darwinism — Darwinism purged of the Lamarckian element that even Darwin himself had retained. But in other respects Weismann's break with the past did not seem quite so decisive. He had rejected Lamarckism, but germinal selection was seen as a concession to equally non-Darwinian ideas. Nor had he made the clean break with the growth analogy that modern Darwinians regard as a consequence of his theory. His last survey of evolutionism (trans. 1904) included chapters on the regeneration of lost organs and on the biogenetic law (Haeckel's name for the recapitulation theory). Weismann admitted that recapitulation was not a direct result of evolution, but he accepted it as a frequent byproduct of the process in which the germ plasm produces new characters (1904, 2:185–87). By his refusal to take a decisive stand against Haeckel on this point, he revealed his own loyalty to the developmental tradition.

If Weismann unpacked the implications of the original form of Darwinism, to most of his contemporaries this seemed only to expose the weakness of the theory's foundations. Far from putting Darwinism on a firmer footing, his campaign backfired and ignited the wave of anti-Darwinian feeling that swept through biology in the decades around 1900. His dogmatism undermined the compromise that had sustained pseudo-Darwinism, but his ideas seemed so speculative that many biologists preferred to move into the neo-Lamarckian camp. In the light of this reaction, surely Weismann cannot be treated as the biologist responsible for completing the Darwinian Revolution. It may be more fruitful to see him not as the founder of modern ideas on heredity but as another aberrant product of the developmental tradition within which Darwin himself had worked. The biologists who revolutionized early-twentieth-century ideas on inheritance and variation may have adopted Weismann's concept of hard heredity, but they were determined to make a far more decisive break with the past. They rejected natural selection altogether, as they rejected all the products of the era in which the analogy between growth and evolution had flourished.

Mendelism revolutionized biologists' ideas about the relationship between growth and evolution because it established a distinction that is taken for granted today but that was ignored in the conceptual framework of nineteenth-century developmentalism. Genetics established itself as a distinct branch of science by divorcing the study of heredity from embryology. The geneticists decided that the question of how characters are transmitted from one generation to the next need not be linked to an investigation of the far more difficult issue of how the characters develop in the individual organism. Mendelian genetics was highly successful in dealing with its more limited topic, and its success created a climate of opinion in which it seemed obvious that individual growth was not a suitable model for evolution. The appearance of evolutionary novelty depends on the origin and transmission of genetical characters, not on how those characters control the course of individual development.

Darwin and Weismann had rejected the explicit teleology of the developmental world view, but they left the link between growth and evolution intact. Thus it is the "Mendelian Revolution" rather than the "Non-Darwinian Revolution" that was the more decisive event in the breakdown of the old tradition. As mentioned above, the term *Mendelian Revolution* need not be taken too seriously; it is used here merely to indicate that major conceptual changes were necessary in the period between the *Origin* debate and the creation of modern evolutionism. The term also suggests that this later revolution took place outside the research traditions that had supported both Darwinism and pseudo-Darwinism. Genetics certainly had major implications for evolutionism, but it was several decades before these implications were acknowledged by the more orthodox biological disciplines. In the end, the Mendelian and Darwinian revolutions were completed side by side, through the emergence of population genetics and the synthetic theory of evolution.

"Mendelian Revolution" is an even less appropriate description than its Darwinian equivalent, since Gregor Mendel himself did not participate in the event that bears his name. He died in obscurity before his laws of inheritance were "rediscovered" in 1900 and became the foundation of modern genetics. There is now considerable doubt as to whether Mendel even intended to pioneer a new approach to heredity. Robert C. Olby (1979, 1985) pointed out that by the standards of the early twentieth century, Mendel was not a "Mendelian." His real concern was to investigate hybridization as an alternative to transmutation in the production of new species. The geneticists who referred to his work actually read a good deal of their own thoughts into Mendel's long-neglected paper. Onno Meijer (1985) suggested that Carl Correns proclaimed Mendel's priority in order to prevent Hugo De Vries from being given all the credit for his independent discovery of the famous 3 to 1 ratio of characters in the second hybrid generation. Mendel may thus have been made the figurehead of the new science in order to stave off a priority dispute among its real founders. Here, as in the case of Darwin, it is necessary for the historian to look behind the facade of the hero-myth created by the scientists for their own interests.

Whatever Mendel's intentions, his paper included a clear demonstration of the laws governing the inheritance of discontinuous variations. The potential value of this demonstration could not be recognized, however, as long as growth was seen as an integral feature of the reproductive process (Sandler and Sandler 1985). Nineteenth-century biologists were simply not prepared to accept a study of inheritance which ignored the question of growth. Modern genetics emerged when a number of biologists at last gave up the study of growth to concentrate on heredity as a distinct field of study (Horder, Witkowski, and Wylie 1986). The new discipline synthesized ideas originating from a wide range of areas including cytology, biochemistry, and breeding studies (for classic surveys of the origins of genetics see Dunn 1965; Sturtevant 1965; Carlson 1966; Stubbe 1972; Olby 1985; useful surveys include Mayr 1982, pt. 3, and Bowler 1984a, chaps. 9, 11).

Genetics represented the triumph of a new breed of experimental biologists who had set out deliberately to transcend the limits of the old morphological program. Determined to extend the experimental techniques so successful in physiology to the study of generation, they first attempted to create a science of developmental mechanics. Embryology would be turned from a purely descriptive into an experimental science through an investigation of the processes that actually built the structure of the growing embryo. Although doomed to failure, this program defined certain issues that were resolved by the geneticists, at least to their own satisfaction. Was the growth of the embryo predetermined by information somehow "coded" into the material of the cell nucleus, or could it be influenced by the cytoplasm, the non-nuclear material of the cell? Could the environment itself determine some aspects of growth? In the end, the geneticists opted for nuclear preformation, although this viewpoint only gained widespread acceptance after a long and bitter debate (Coleman 1965; Churchill 1970; Gilbert 1978; Maienschein 1984, 1986).

The new breed of embryologists had no patience with the old morphological approach or with the tacit assumption that the purpose of embryology was to throw light on evolution via the recapitulation theory. Was there, then, a "revolt against morphology" at the end of the nineteenth century (Allen 1975a)? Recent research has uncovered a process of development *within* traditional disciplines that only gradually led to a rejection of the old way of thinking (Maienschein 1981; Bensen 1981). Several early geneticists began as embryologists working with the recapitulation theory, but they gradually changed their orientation as they tried to accommodate the new interest in developmental mechanics. Eventually attention began to focus not so much on the study of growth, which proved extremely difficult with the techniques then available, but on the transmission of characters from parents to offspring. Once heredity was defined as a legitimate area of study, independent of growth itself, the stage was set for the creation of a science that would see evolution purely in terms of the appearance and spread of new genetical characters. The production of those characters by mutation would be governed not by the growth-process but by events taking place within the genetical material itself. Here at last were the foundations of a new approach to evolutionism that would repudiate all the non-Darwinian aspects of the nineteenth-century developmental viewpoint.

Did Darwinism itself play a role in the creation of this new approach? Weismann's theory of the germ plasm was certainly part of the ferment of ideas out of which the new science was created, but in general the early geneticists wanted nothing to do with Darwinism. They had turned their backs on the recapitulation theory and were thus intrinsically hostile to the suggestion that what they were doing should be interpreted in the light of the evolutionary past. They rejected the pseudo-Darwinism of Haeckel *and* the use of the recapitulation theory by anti-Darwinians such as the American neo-Lamarckians. Inevitably, perhaps, it was Haeckel's "Darwinism" that was most often singled out for criticism, since this symbolized the earlier generation's commitment to evolutionary issues. In addition, though, the geneticists became convinced that adaptation could not play the role attributed to it by Darwin (and by the neo-Lamarckians). Whatever the process by which new characters are produced, it was considered internal to the organism and not under the control of the external environment. Biologists whose chief concern was laboratory work could all too easily ignore lessons that had been learned by field naturalists working with animals and plants in the wild. The possibility that the environment might limit the *spread* of a new character through a population was, as yet, of no interest to the geneticists, since they studied only the transmission of characters in groups of organisms isolated under artificial conditions. In such a climate of opinion, the anti-Darwinian, that is, anti-adaptationist, arguments used by the supporters of orthogenesis could be exploited in an entirely new context.

This anti-Darwinian tendency can be illustrated through the work of two leading geneticists, William Bateson and Thomas Hunt

Morgan. Bateson started out as an evolutionary morphologist working on the origin of the vertebrates, but he became dissatisfied when he realized that his hypothetical ancestry was likely to remain forever untested by fossil evidence. In his *Materials for the Study of Variation* (1894) he argued that a new approach to evolution was needed, concentrating on the direct evidence for the production of new characters. He was now convinced that evolutionary novelties are produced by saltation, not by the natural selection of individual variation. Some internal, purely biological process must be responsible, and Bateson argued against the claim that adaptation plays a significant role in evolution. Morgan began his career as an embryologist, but he, too, soon began to advocate saltation as the only source of new characters (Allen 1968, 1969, 1978). His *Evolution and Adaptation* (1903) was a bitter critique of the Darwinian view that adaptation is the driving force of evolution. Here, in the early work of two biologists who would go on to play major roles in the foundation of modern genetics, is the continued use of anti-Darwinian arguments of the kind that can be traced right back to Mivart (Bowler 1983, chap. 8). Like Mivart, Bateson and Morgan wanted to believe that internal forces direct evolution along predetermined lines. Their saltatory theory was an effort to suggest that evolution could become accessible to experimental study through direct observation of how new characters appear. Both at first believed that internal factors would determine the kind of new characters which would be produced, giving a new foundation to the old idea of orthogenesis.

One of the the few figures in this story who had a genuine sympathy with Darwinism was the Dutch botanist Hugo De Vries. He deliberately called his first theory of heredity "intracellular pangenesis" (1889, trans. 1910a), although it bore only a superficial resemblance to Darwin's pangenesis. De Vries later produced what many biologists saw as the most convincing evidence that new characters originate by saltation. He found what appeared to be new varieties or subspecies formed by sudden transformations within the evening primrose, *Oenothera lamarckiana*. On the basis of this evidence De Vries created his "mutation theory" (1901–03, trans. 1910b), arguing that new varieties are produced not by the natural selection of individual variation but by the sudden appearance of mutated forms that immediately breed true. For T. H. Morgan and many others, this confirmed that Darwin had been wrong, but De Vries himself insisted that natural selection must still work at a different level (Allen 1969; Bowler 1978). The mutated varieties were not produced in any consistent direction, so they could be seen as a new form of random variation, with only the fittest surviving beyond a few generations. In De Vries's view, the basic course of evolution would thus still be determined by adaptation.

De Vries played an important, if transitory, role in the event that is now hailed as the key to the emergence of the new science of genetics. Along with another botanist, Carl Correns, he "rediscovered" the work of Gregor Mendel, the abbot whose studies of hybridization during the 1860s are seen as anticipations of the modern laws of heredity. Mendel's experiments with peas showed that parental characters do not blend but are transmitted unchanged from one generation to the next. Certain "recessive" characters can be masked by their "dominant" equivalents or alleles, but they can reappear unchanged in later generations, as illustrated by Mendel's classic 3 to 1 ratio in the second hybrid generation of his peas. To many of the early geneticists, it seemed that Mendel's laws could only be interpreted as evidence that the particles responsible for coding the genetic information come in pairs, each parent contributing one of its own pair to the offspring. The 3 to 1 ratio was noticed independently by De Vries and Correns in 1900, and soon Mendel's laws were being hailed as the basis for a new understanding of heredity. Bateson (1902) provided an English translation of Mendel's original papers, along with a powerful defense of the new initiative.

Mendelism developed rapidly during the early decades of the new century. Yet, with the exception of De Vries (who soon lost interest), few converts saw the new theory as a solution to the problems of Darwinism. By opting for saltation instead of natural selection, Bateson had already made an enemy of Pearson. Mendelism thus almost inevitably became caught up in the dispute between Bateson and the biometrical school (Cock 1973). The geneticists studied only discontinuous variation of the kind displayed by Mendel's peas, dismissing the continuous range of individual variation studied by the biometricians as of no genetical or evolutionary significance. The Danish biologist Wilhelm Johannsen showed that when applied to the existing range of variation, selection always seems to reach a limit. To transcend this limit, mutations would be required to create entirely new genetical characters. De Vries's mutation theory was increasingly modified along lines more familiar to the modern geneticist: it was accepted that mutations do not create new subspecies, but feed additional genetical characters into the existing population, thereby extending its variability. The majority of geneticists still could not accept De Vries's view that survival value can influence the spread of a mutated character. They assumed that evolution must be driven by "mutation-pressure." If the same mutation appears repeatedly within the population, the species will be forced to evolve in the corresponding direction, whether or not the new character is of any use in the struggle for existence.

In America, T. H. Morgan at first remained suspicious of Mendelism. He was finally converted in 1910, when his own experiments showed that the chromosomes in the cell nucleus can be interpreted as the physical site of particles responsible for the transmission of fixed characters according to Mendel's laws. Working with the fruit fly, *Drosophila*, Morgan and his associates undertook a series of experiments revealing the link between the chromosomal theory of inheritance and Mendelism (Morgan et al. 1915). Genetics was confirmed as a distinct area of biology, and its leaders set out to create an academic power-base centered on the new discipline (Sapp 1983). The difficult problems of developmental mechanics were pushed aside so that the geneticists could concentrate on the highly successful study of transmission. At least in part, the severing of the link with embryology was a deliberate move by which geneticists such as Morgan made a bid for the research funding that would become available through the application of their techniques to animal and plant breeding (Allen 1986a). Soon, however, the new science established itself within the American university system and could thus develop independently of its practical applications.

Scientists generally assume that a new discipline arises automatically on the foundations of any great discovery. But as Jan Sapp (1983, 1987) has shown, the science of genetics was to some extent *constructed* by biologists seeking to further their own careers. The scientific revolution was not a purely conceptual affair; it also involved a social process by which the new field was defined as an independent area of study with a power-base of its own within the scientific community. The Mendelian Revolution — with all its consequences for evolution theory — was a product of political maneuvering within the American and British university systems. The clearest evidence that this social factor had a positive bearing on the kind of science that was produced can be seen in the fact that the revolution did not take place in the same clear-cut manner in Germany. The American geneticists built their new paradigm around the assumption that the chromosomes are the sole bearers of heredity. The cytoplasm surrounding the cell nucleus was dismissed as irrelevant, and any effort to show that it might play even a subsidiary role in inheritance was rigorously suppressed. The German biologists refused to accept this rigid distinction and allowed studies of cytoplasmic inheritance to flourish along lines that seemed outlandish to English-speaking geneticists (Harwood 1984, 1985).

The issue of chromosomal versus cytoplasmic inheritance is of particular importance in assessing the impact of the Mendelian Revolution on evolution theory. The chromosomal theory encouraged a rigid definition of the gene as a physical unit responsible for transmitting the information by which a particular character of the adult organism is built. Because these units were seen as distinct entities encoding the information by some (presumably biochemical) means, the character itself was regarded as a unit that must be transmitted unchanged from parent to offspring. Only mutation could change a gene, producing a new character through some rearrangement of the physical structure responsible for coding the information. The organism's genetic endowment (its *genotype*, to use the term coined by Johannsen) had to be viewed as quite distinct from its physical form (its *phenotype*), because of the presence of recessive genes. Evolution thus necessarily became a process of genetical, not phenotypic, change. In such a system the analogy between growth and evolution was irrelevant, and Lamarckism was inconceivable. Variation was always a disturbance of, not a continuation of, growth, and characters acquired by the adult organism simply could not be translated into equivalent modifications of the genotype. Mendelism thus not only set itself up as a rival to the old morphological sciences but also undermined the conceptual foundations upon which those older sciences had built their non-Darwinian view of evolution.

The possibility that inheritance might be transmitted by the cytoplasm as well as by the nucleus threatened to limit the impact of the Mendelian Revolution upon evolution theory. If the cytoplasm played a role, then changes affecting the organism as it grows might, after all, be able to play a role in evolution. Cytoplasmic inheritance would keep alive the hopes of Lamarckians and all those who wished to retain the analogy between growth and evolution. Small wonder that the English-speaking geneticists refused to countenance either cytoplasmic inheritance or Lamarckism, since both would diminish the extent to which their new science could be presented as a distinct entity. They not only suppressed the study of cytoplasmic inheritance but also discredited those who still tried to provide evidence of Lamarckism. When the Austrian biologist Paul Kammerer tried to interest British and American geneticists in his Lamarckian experiments with the "midwife toad," Bateson led a move to brand him as a charlatan (Koestler 1971). Similar efforts by British and American scientists to present Lamarckism as an experimental fact were ignored because without better knowledge of the cytoplasm they had no explanation of how the acquired characters might be inherited. Whatever the moral or philosophical advantages of Lamarckism, the theory soon began to wane once the indirect support from the analogy with growth was cut off.

Nevertheless, it is important to note that the rigid elimination of non-Mendelian theories was very much a product of the English-speaking scientific community. Whatever Kammerer's status as a scientist, cytoplasmic inheritance and non-Mendelian evolutionary mechanisms continued to enjoy the support of many German-speaking biologists. Where the cytoplasm was still taken seriously, not only Lamarckism but also the link between evolution and growth retained some plausibility. One consequence of this was that German biology did not undergo the fragmentation so obvious in Britain and America. Those biologists working in the morphological sciences did not feel completely alienated from the new experimental disciplines. A paleontologist such as Otto Schindewolf could interact freely with geneticists in an attempt to create a synthesis based on non-Darwinian evolutionary mechanisms (Reif 1983, 1985). English-speaking biologists thus experienced a far more drastic Mendelian revolution than their German colleagues, with a correspondingly more complete elimination of non-Mendelian — and non-Darwinian — evolutionary ideas. While the Germans were still trying to salvage some aspects of the old morphological tradition, British and American biologists were forced to confront a situation in which the older non-Darwinian mechanisms were unacceptable. Their search for an evolutionary theory that would be compatible with the more hereditarian position led eventually to the Modern Synthesis.

The Modern Synthesis

English-speaking paleontologists and field naturalists certainly did not welcome the Mendelians' rejection of Lamarckism and the growth analogy. For several decades they simply ignored the new genetics, leaving biology fragmented in a way that cried out for a new initiative to unite the orthodox and radical branches of the science. But where could such a synthesis come from? The geneticists had ruled out not only Lamarckism but any significant role for adaptation in evolution. Yet field naturalists and, to a lesser extent, paleontologists were by now convinced that adaptation to the environment must be a major factor. Admittedly, some paleontologists still opted for nonadaptive orthogenesis, and at first it looked as though the geneticists' concept of mutation pressure might provide a viable mechanism for directed evolution. But the evidence for the random nature of mutations soon ruled this out. It now became obvious that all sides would have to adopt a more flexible posture if any new initiative was to succeed.

The modern synthetic theory of evolution drew together all of the strands sketched in above (for surveys see Mayr and Provine 1980; Grene 1983; Bowler 1984a, chap. 11). Although normally regarded as a re-emergence of Darwinism, it should now be apparent that at least one component of the synthesis came from a new science owing no allegiance to nineteenth-century traditions. But the geneticists were now left with a difficult choice: if evolution was not to be totally random, the spread of new mutations would have to be controlled by a selecting agent in the external environment. Gradually they began to concede that adaptation must play a role in determining the success or failure of a mutated gene in a wild population. They now needed to study the genetical structure of large populations, not small laboratory samples. To do this they needed the experience of the biometricians, who had rejected Mendelism in favor of the statistical analysis of large-scale variation. Population genetics was created by synthesizing the two most revolutionary biological developments of the turn of the century — Mendelism's particulate theory of heredity and the biometricians' sophisticated populational approach (Provine 1971). To the extent that Pearson had explored concepts implicit in Darwin's original selection theory, population genetics drew upon the Darwinian tradition, and the genetical theory of natural selection represents a completion of Darwin's program.

Yet even biometry had to some extent cut the selection theory off from its roots, as is evident from the fact that several early exponents of population genetics ignored the characteristic Darwinian concern for migration and the role that geography must play in evolution. Ernst Mayr (1959c, 1976) argued that the field naturalists' contributions to the synthesis should not be overlooked in the rush to emphasize the revolutionary aspects of population genetics. Mayr is quite right to emphasize that field studies provided the most genuinely Darwinian element in the synthesis, an element all too easily ignored in some more abstract expositions of the genetical selection theory. But Provine (1978) is probably correct to insist that the more familiar developments in population genetics should be given pride of place. Without the synthesis of genetics and the selection concept, it seems unlikely that the field naturalists' studies of biogeography and speciation would have served to convert the scientific community to Darwinism. In the end, the traditional disciplines had to adapt themselves to what had now emerged as the only plausible evolutionary mechanism, although they certainly helped to show how that mechanism should be applied to the real world.

American and British geneticists assumed that mutation was the only source of evolutionary novelty, and at first they continued to argue that adaptation was not a significant factor in evolution. Only those effects accessible to study in the laboratory were to be allowed in evolution theory; in effect evolution was to become a scaled-up version of a laboratory breeding program. Paleontologists and field naturalists cannot be blamed for refusing to take such a one-sided view seriously. They knew that adaptation *was* important, and even the support for orthogenesis was based on the belief that evolution is a directed process, not a summation of random genetical mutations. Some early geneticists, including T. H. Morgan, had toyed with the possibility that mutations might occur consistently in a particular direction, thus forcing evolution along an orthogenetic track. But Morgan's own experiments with *Drosophila* revealed a bewildering variety of mutations, effectively eliminating the hope of explaining orthogenesis by mutation-pressure. It now seemed clear that mutation is a disturbance of the genetical material by outside factors such as radiation, and there seemed little evidence that the structure of the genetical material might somehow limit the disturbances to a particular direction. Mutations simply extend the range of a population's variability in all, or at least a great many, directions. In these circumstances, evolution directed by mutation alone would become a totally haphazard affair that could not even guarantee the continued existence of recognizable species.

Morgan himself soon decided (1916) that something must control the direction of evolution, and he conceded that only those mutations conferring adaptive benefit would spread into the population. The path toward a reconciliation with Darwinism now began to open up; in the new Mendelian climate of opinion, survival determined by adaptive fitness remained as the only way in which random genetic mutation could be constrained and directed. Despite their initial hostility, the geneticists were forced to adopt natural selection or leave evolution as a totally undirected process. As yet, though, Morgan's concession did not allow selection to be seen as an effect operating continually upon the whole population. He refused to admit that — because many mutated genes are recessive — selection is unable to prevent deleterious characters from being reproduced. To develop a viable theory of adaptive evolution in the wild, the Mendelians needed a technique for studying the genetical composition of the whole population and the effect of differential reproduction on that composition.

A science of population genetics was required, and this would necessarily have to be statistical in character. Statistical techniques had, of course, been developed by the biometrical school, but Pearson's hostility to Mendelism had ensured that no attempt would be made to study the genetical structure of whole populations. As support for

Mendelism grew, however, Pearson's followers found it increasingly difficult to sustain their leader's hostility. In 1918 Ronald Aylmer Fisher began the process of applying biometrical techniques to the study of populations whose inheritance was governed by Mendelian factors. It soon became clear that if a number of different genes influence a single character (such as the height of the individual), then their combined effect on the whole population will be to produce the continuous range of variation that Pearson and the Darwinians had always seen as the raw material of selection. Continuous variation and discontinuous mutations are *not* incompatible once it is recognized that the genetical structure of a wild population is far more complex than the artificially restricted samples studied by the early geneticists. Along with J. B. S. Haldane, Fisher went on to create the basis for a theory of population genetics that would allow natural selection to maintain the genetical balance of a large population, whatever the input from mutation. The balance could then be changed quite rapidly by differential reproduction if one or more genes became favored in a new environment (Fisher 1930; Haldane 1932; see Provine 1971). Something resembling the original Darwinian view of natural selection now began to seem the most plausible mechanism of evolution compatible with the Mendelian model of inheritance.

Fisher and Haldane stressed the extent to which their work helped to link the Darwinian and Mendelian traditions. Fisher explicitly presented genetics as the solution to Darwin's problems over the nature of heredity and argued that the Mendelian interpretation of particulate inheritance was more or less self-evident (1930, p. ix and chap. 1). He thus helped to create the "jigsaw puzzle" model for the growth of evolutionism subsequently taken up by Eiseley (1958) and other historians. Yet there was one area in which Fisher and Haldane were still failing to address the issues that concerned the field naturalists who were the more direct heirs of Darwin's biogeographical research program. Both saw natural selection as a process acting uniformly on a large population exposed to a new environment across its whole range. There was no effort to investigate what might happen when small populations became isolated from the main body and no real interest in the problem of speciation. In America, meanwhile, a different form of population genetics had been worked out by Sewall Wright (1930, 1931; see Provine 1986). Drawing inspiration from his work on artificial selection, Wright focused on small populations among which inbreeding might create new interaction systems between genes through "genetic drift." In Wright's theory, the most effective situation for evolution would be a large population imperfectly broken up into local groups. Drift would allow the small populations to make rapid transitions to new adaptive states that would then spread through the whole species. Significantly, Wright modified his original theory to include a role for nonadaptive evolution, in accordance with the prevailing assumption that some minor characters are useless — a view gradually abandoned once the new Darwinism began to gain momentum.

Wright's theory was closer in spirit to the work of the field naturalists. Here the study of geographical distribution had maintained the focus of attention on the ability of local populations to adapt to new territories into which they had migrated. Bernhard Rensch (1980) described how his own initial preference for Lamarckism as the mechanism of local adaptation was undermined in the late 1930s as he became aware of the genetical theory of natural selection. Ernst Mayr (1959; Mayr and Provine 1980), however, argued that many field workers adopted a populational (and hence Darwinian) view of species without any awareness of population genetics. Both sides were, in effect, converging on the same solution. Population genetics, especially in the form proposed by Wright, now seemed the appropriate foundation upon which to build a theory compatible with the naturalists' interests. Theodosius Dobzhansky's *Genetics and the Origin of Species* (1937) played a vital role in translating the abstruse mathematics of population genetics into propositions that made sense to the field workers. The emergence of the new Darwinism was proclaimed in books such as Mayr's *Systematics and the Origin of Species* (1942) and Julian Huxley's *Evolution: the Modern Synthesis* (1942). George Gaylord Simpson's *Tempo and Mode in Evolution* (1944), by re-evaluating the evidence that had previously been interpreted as favorable to linear rather than divergent evolution, helped to convince paleontologists that the new synthesis should be taken seriously.

Initially the new theory still allowed room for some non-Darwinian effects. Wright's genetic drift and Simpson's "quantum evolution" both seemed to allow small populations to pass through a temporary nonadaptive phase to reach a new ecological niche. Stephen Gould (1980, 1983) pointed out that the synthesis soon "hardened" into a rigid adaptationism that excluded such non-Darwinian phases. Darwin's view of evolution thus enjoyed a belated triumph, at least in the English-speaking scientific community. The geneticists' suspicion of the link between growth and evolution had now spread throughout modern biology. Mendelism had succeeded where Darwin had failed, but it had unexpectedly revealed that some of Darwin's insights had been valid after all. The Darwinian Revolution was now complete, although the synthetic theory can hardly be seen as the inevitable end product of the process that Darwin had set in motion.

The Darwin industry in the history of science was established in the 1960s, when the synthetic theory was at the height of its popularity. Indeed, the theory became so powerful that critics outside science accused biologists of setting up a dogmatic orthodoxy that stifled dissent. In recent decades there has been an intensification of debate, although the new ideas presented have often been modifications of, not alternatives to, the synthesis (Cherfas 1982; Smith 1982; Ruse 1982). The best known of the new concepts is Niles Eldredge and Stephen Gould's theory of "punctuated equilibrium," which posits that isolated populations diverge rapidly from the norm for their species, and successful modifications later supplant the parent form over its whole territory. This theory has been introduced to solve the persistent problem of the discontinuity of the fossil record, since it allows for the apparently abrupt changes that seem to punctuate so many fossil sequences. Eldredge and Gould are suspicious of Darwin's claim that the imperfection of the record hides the gradual character of all evolutionary change. Significantly, Eldredge (1986) has argued that in light of the record's obvious discontinuity, Darwin's commitment to gradualism should be seen as something of a historical puzzle. The orthodox Darwin industry sees gradualism as a more or less obvious aspect of evolutionism, but a scientific dissident perceives it as an anomaly that needs to be explained. Eldredge's position reveals the extent to which the perception of science's past is shaped by modern interests and points to the need for a more general revaluation of Darwin's position in the growth of modern evolutionism.

Because the Darwin industry has followed Fisher's assumption that genetics merely fills in the gaps in Darwin's thinking, most accounts of the Darwinian Revolution have ignored the role played by the non-Darwinian theories of the late nineteenth century. If there was no real conceptual revolution in biology around 1900, it has seemed reasonable to assume that the Origin was sufficient to force everyone to confront the implications not just of evolutionism but of Darwinian evolutionism. The alternative account of the history of evolutionism sketched in above must inevitably throw doubts on the traditional interpretation of Darwin's impact on nineteenth-century thought. Western culture obviously had to accommodate the idea of evolution once the Origin had forced scientists to take it seriously. But if few scientists accepted the Darwinian view of evolution, is it still possible to believe that everyone else was profoundly affected by his more radical ideas? The fact that most scientific evolutionism was at first non-Darwinian in character suggests that the theory's application to human affairs may also have followed the developmental model.

Chapter Six

HUMAN EVOLUTION

✠ As soon as it became clear that the scientific world had endorsed the theory of evolution, Western culture as a whole was forced to begin a search for new foundations upon which to build its moral and social values. As Chambers had indicated in his *Vestiges of Creation* — and as Darwin had known all along — an evolutionary interpretation of the history of life on the earth must inevitably extend itself to include the origins of the human race. Humanity could no longer regard itself as a direct product of the Creator's power, and it became difficult, if not impossible, to retain the old transcendental source of moral values based on divine revelation. As the evolutionists threw themselves into the task of reconstructing the course of life's development, it was obvious that an account of how the human body had originated was not enough. If the body had evolved from a lower form, then so had the mind, and so had human moral and social values. Perhaps evolutionism itself would provide the new foundation for morality, since a study of how human beings had evolved in the past might tell them how they should behave in the present.

The transition to a naturalistic source of ethical values was inevitably traumatic for those who still cherished the old Christian virtues. The Darwin industry has certainly sought to exploit the sense of crisis invoked by this revaluation of *Man's Place in Nature*, to quote the title of T. H. Huxley's classic book (1863; Huxley 1893–94, vol. 7; see Young 1973, 1985a). No account of the Darwinian Revolution is complete without a discussion of the *Descent of Man* and the objections to Darwin's naturalistic explanation of human faculties by a host of conservative thinkers, including (on this issue) Lyell and Wallace. Yet this concentration on the traumatic aspects of the debate seems rather shortsighted when measured against the general sweep of nineteenth-century thought. Once the sense of confrontation died away, the foundations of a new morality were soon established, and the structure built upon them was by no means as radical as many conservatives had feared. Evolutionism did not usher in an age of moral nihilism, despite the complaints of "social Darwinism" that so confused a later generation of historians. Almost everyone was determined to ensure that the human race and Western culture would both retain their positions at the center of the universal scheme of things. This resolution was implemented by insisting that evolution must be purposeful and, hence, progressive. The human race was important because it was at the forefront of nature's steady march toward a higher state. Furthermore, the fact that nature *was* progressive ensured that the lessons it taught were indeed moral lessons, designed to enhance the perfection of the human race.

The extensive use of non-Darwinian models of evolution by late-nineteenth-century biologists reflects this preference for a more teleological approach. I have already argued that the popularity of the analogy between growth and evolution suggests that the biogeographical and adaptationist aspects of Darwin's theory did not play a major role in shaping biological thought. In generalizing this study to include the broader implications of evolutionism, it becomes necessary to move outside biology to see if other areas were able to promote a developmental viewpoint. If there were nonbiological sources of evolutionism, and if those sources were also promoting a non-Darwinian view, then there is yet another line of evidence suggesting that Darwin's position in the history of evolutionism must be reassessed.

Even the physical sciences could provide support for progressionist evolutionism. Ronald Numbers (1977) and Stephen Brush (1987) have emphasized the extent to which the nebular hypothesis in cosmology paved the way for an evolutionary world view. Originally proposed in the eighteenth century by Immanuel Kant and Pierre Simon de Laplace, the nebular hypothesis supposed that the solar system has condensed from a vast cloud of dust under the influence of gravity. Many nineteenth-century scientists saw this as the most likely physical explanation of planetary origins. Both Robert Chambers and Herbert Spencer built the nebular hypothesis into the foundations of their evolutionary philosophies. A developmental account of the earth's origin paved the way for a similarly progressionist interpretation of the history of life on the earth. This image of cosmological evolution was distinctly non-Darwinian in character. It portrayed development as a process grinding inexorably on toward a predictable goal, with no room for haphazard divergence in response to changing conditions.

More directly related to my present purpose is the independent origin of an evolutionary view of human society. As the sciences of sociology, anthropology, and archaeology emerged in the mid-nineteenth century, all adopted an essentially developmental view of history. At one time, historians simply assumed that this movement was triggered by the Darwinian Revolution in biology. The Darwin industry, naturally enough, has not sought to challenge an assumption that serves its own interests so well. Few accounts of the Darwinian Revolution spend much time on contemporary developments in anthropology or prehistoric archaeology. If social evolutionism is mentioned, Herbert Spencer's work is invariably taken as the key example — an example that quite conveniently fits the Darwinian model. Ostensibly, this indifference to the work of the anthropologists and archaeologists is justified by the fact that many of them did not discuss the biological origins of the human race. But this highly selective approach to the topic of human evolutionism has allowed the Darwin industry to insulate itself from current trends in the history of fields such as anthropology, trends that generally confirm the non-Darwinian origins of the developmental view of society. Enshrined in a cosy world defined by the debate on "man's place in nature," historians of evolution theory have failed to notice that other areas of discourse were already shaping a progressionist model that would have far more impact on the evolutionary world view.

Anthropology and Evolution

In a wide-ranging survey of nineteenth-century thought, Maurice Mandelbaum (1971, pt. 2) presented social, anthropological, and biological evolutionism as parts of the same broad movement. The essential character of this movement was its commitment to *historicism*, or the assumption that a study of past developments will allow an understanding of the future, because a fixed pattern of progress is somehow built into the very nature of things (see also Popper 1957, chap. 2). Robert A. Nisbet (1969, chap. 5) gave a detailed outline of the central principles of nineteenth-century social evolutionism. They include the belief that change is natural, directional, and continuous, all of which could presumably be reconciled with a Darwinian model of branching evolution. But change was also assumed to be immanent and necessary, that is, to proceed inevitably from the very nature of the entity whose evolution is being studied. As Nisbet pointed out, these attributes are much more difficult to reconcile with a theory of purely adaptive evolution. More generally, the historicists' claim that they had discovered predictive laws of development seems to rule out any theory of "haphazard" evolution, or indeed any theory that permits the unexpected to happen. The stages of development *must* succeed one another in their inevitable sequence, a view that seems to dovetail most naturally with the non-Darwinian mechanisms of biological evolution, especially those modeled on the analogy with individual growth.

These characteristics are, of course, meant to apply to the whole spectrum of nineteenth-century social evolutionism, including the theories of Comte, Hegel, and Marx. In the English-speaking world, the most clearly defined form of historicism can be found in the work of anthropologists such as Edward B. Tylor and Lewis H. Morgan, who saw the social evolution of each race as the ascent of a more or less rigidly preordained hierarchy of cultural stages. Archaeologists, too, seem to have preferred a linear image of cultural development. Biological evolutionism was eventually used to link the idea of cultural progress with pre-existing beliefs about the relative mental ability of the various races, so that the "brighter" races were supposed to have advanced furthest from the ape, and furthest up the cultural scale. Unlike Herbert Spencer, who admitted, at least in principle, the divergent character of social and biological evolution, the anthropologists were committed to the idea of a rigid sequence of developments. Even when they did not engage directly in speculation about the biological origins of the human race, their theory provided an obvious model upon which to base a non-Darwinian image of progress along a linear scale.

The assumption that the rise of interest in cultural evolutionism was initiated by the advent of Darwinism was presumably based on the coincidental timing. Since the *Origin of Species* was published in 1859, and cultural evolutionism flourished in the 1860s and 1870s, the possibility that the debate in biology had stirred the anthropologists' imagination was at least plausible. One leading survey of the history of archaeology still accepts that Darwinism played a role in promoting interest in the primitive state of early humans (Daniel 1975, 63–66). Yet the general trend among historians of anthropology and archaeology has been to repudiate this link. Cultural evolutionism certainly flourished in the same environment that generated concern with the history of life on the earth. But the evidence suggests that the exploration of this theme by the human sciences was already underway before Darwin was published, and that it was influenced by factors that were only peripherally linked to Darwin's biology (Burrow 1966; Harris 1968, e.g., 142–43; Stocking 1968, chap. 5; Hatch 1973, chap. 2). More recently, Stocking (1987) accepted that the Darwinian debate helped to stimulate the emergence of a developmental approach to culture, although he conceded that the archaeologists' simultaneous demonstration of human antiquity played a similar role. Historians of archaeology have now repudiated the once-popular claim that Darwinism precipitated the sudden revolution in attitudes toward prehistory that took place in the 1860s (Gruber 1964; Freeland 1983; Grayson 1983, 211).

One of the earliest accounts of cultural evolution was Sir Henry Maine's *Ancient Law* of 1861, which sought to explain how modern legal systems had arisen from the primitive customs of patriarchal family groups. Edward B. Tylor's *Researches into the Early History of Mankind* appeared in 1865, but was based on work done in the pre-Darwinian era. Tylor's efforts to explain the development of religious and moral customs provide one of the clearest examples of linear progressionism. They were reinforced by John Lubbock's *Origin of Civilization* of 1870 and Lewis H. Morgan's *Ancient Society* of 1877, the latter based on studies of the American Indians (Kuper 1985). Tylor, Lubbock, and Morgan studied existing cultures, but the titles of their books indicate that these cultures are "primitive" in the sense that they are taken to illustrate the earlier states of "higher" societies. A similar approach to social development was promoted in Germany by Johann Bachofen and Adolf Bastian. The movement flourished into the last decade of the century; James Frazer's celebrated *Golden Bough* (first published in 1890) was one of its last major expressions.

As Burrow (1966) explained, the British evolutionary anthropologists developed their system in order to reconcile their belief in the constancy of human nature with the evident diversity of cultures around the world. In some cases, efforts had been made to educate colonized peoples so that they would accept the "obvious" superiority of European cultural values. The failure of these efforts convinced the anthropologists that there must be a fixed sequence of stages through which all cultures develop. It was thus impossible artificially to transfer a culture from one stage to another; the development had to take into account the internal logic of cultural progress. The "psychic unity" of mankind ensured that all races tend to progress in the same direction, along the same hierarchy of cultural stages. It was Morgan who defined the major stages as savagery, barbarism, and civilization, arguing that they formed a "natural as well as a necessary sequence of progress" guided by a "natural logic which formed an essential attribute of the brain itself " (1877, 11, 59). He believed that all but the lowest stages of social development were still visible in the world.

Since cultural development follows its own internal pattern of progress, the anthropologists' argument went, geographical factors have little to do with cultural diversity. Geographically separated peoples do not tend to develop their own unique culture, and hence there can be no significant divergence in cultural evolution. Geographical isolation can only ensure that different races develop at different rates, but all must pass through the same sequence of stages. Similarities between remote cultures are due to the independent invention of the same tools and customs at appropriate points on the scale. The range of societies around the world displays, in effect, a *historical* sequence: the Europeans have advanced furthest along the scale of development, while other races, advancing at a slower pace, reveal the earlier stages through which the Europeans must have passed in the course of their upward progress. The "comparative method" thus allowed the European anthropologist to use modern "primitives" as illustrations of how his own ancestors had lived in the past. In Tylor's view, anthropology was a reforming science because it encouraged the exposure of "survivals" of ancient cultural stages that have still not been eliminated from more advanced societies.

The analogy between this concept of linear cultural development and the non-Darwinian model of biological evolution outlined in earlier chapters should be obvious. Here, in biological terms, is a theory of parallel evolution, or orthogenesis, in which various lines of development advance independently through the same predetermined sequence of stages toward a definite goal. There is no room for any but the most trivial form of divergent evolution in the face of differing environmental conditions. The cultural evolutionists also recognized the similarity between their concept of development and the growth of the individual toward maturity. Tylor himself saw savages as having a mentality equivalent to that of the children of civilized races ([1865] 1870, 108). Frazer made the link with the recapitulation theory even more explicit:

> For by comparison with civilized man the savage represents an arrested or rather retarded stage of social development, and an examination of his customs and beliefs accordingly supplies the same sort of evidence of the evolution of the human mind that an examination of the embryo supplies of the evolution of the human body. (1913, 162)

The assumption that the primitive mind is essentially childlike ran through much late-nineteenth-century anthropological and psychological thought, indicating the widespread popularity of the analogy between evolution and the teleological process of individual growth (Gould 1977b, chap. 5; Muschinske 1977).

Tylor and his fellow anthropologists were well aware that the archaeologists had established a case for the very primitive state of the earliest human societies. In the early nineteenth century, Scandinavian archaeologists such as Christian Thomsen and J. J. A. Worsaae had defined the "three age system" of development from the use of stone tools to those made of bronze and, finally, iron. At first, no one would accept the possibility that the "stone age" might extend so far back in time that the earliest human beings had co-existed with now extinct animals such as the mammoth. When Boucher des Perthes discovered stone tools along with the bones of extinct animals in the gravel beds of northern France, his descriptions were ignored or ridiculed. Around 1860, however, there was a rapid change of opinion in favor of the much greater antiquity of the human race (Grayson 1983). Geologists such as Joseph Prestwich and Charles Lyell now endorsed Boucher des Perthes's work and soon compiled a wide range of evidence for the existence of early humans, summarized in Lyell's *Geological Evidences of the Antiquity of Man* (1863). Lyell certainly discussed Darwin's theory in this book, and his remarks are routinely examined as a contribution to the debate over the *Origin*. But there is little to suggest that Lyell — or any other geologist or archaeologist — accepted human antiquity primarily in response to Darwin's theory.

The primitive state of the earliest human technology was taken as evidence for an equally primitive state of social and cultural development. John Lubbock led the campaign to establish the case for a general progress of society from savage origins, in opposition to those conservative thinkers who still saw mankind as spiritually degenerate (Gillespie 1977). Lubbock's *Prehistoric Times* (1865) coined the terms *Paleolithic* and *Neolithic* to define the advance from the Old to the New Stone Age, indicated by the transition from chipped to polished stone. He argued that modern primitives illustrate the kind of lives led by man's Paleolithic ancestors. By the late 1860s, archaeologists had begun to recognize different tool-making cultures in the Paleolithic. A few (e.g., Dawkins 1874) saw these cultural differences as mere geographical variations, but Gabriel de Mortillet (1883) championed the far more popular assumption that the cultures must represent an evolutionary sequence. In de Mortillet's system, all the inhabitants of Europe had advanced step by step through the same hierarchy of cultural stages toward the Neolithic. This interpretation exactly duplicated the parallelism of the anthropologists' view of cultural evolution. For de Mortillet, the steady progress of human society in the past was a tool to be used in the campaign for further social progress in the present (Hammond 1980). The assumption that all races of mankind must have advanced in the same way was thus built into the conceptual foundations of Paleolithic archaeology.

The concept of a predictable sequence of social development was by no means confined to the anthropologists. In a very different context it occurred in Marx's dialectical theory of social evolution. Other influential thinkers took a rather less rigidly structured view of progress, at least in theory. Herbert Spencer's social philosophy has attracted the most attention from the Darwin industry and certainly seems easier to link with the Darwinian image of branching evolution. At an early stage in his career, Spencer was introduced to K. E. von Baer's view that the growth of the embryo can best be understood as a process of specialization, not as the ascent of a linear scale. As a result, Spencer incorporated the idea of multiple branches of specialization into his evolutionary system. In his *Principles of Sociology,* he argued that modern primitives have probably degenerated from a more advanced state and thus cannot be taken as illustrations of earlier stages in European cultural evolution (1876–96, 1:61, 106). Since each race has evolved in its own environment, each has developed its own unique instincts; thus, "like other kinds of progress, social progress is not linear but divergent and re-divergent" (ibid., 3:325). Against this must be set Spencer's conviction that European society was progressing steadily from militarism to laissez-faire individualism, and that the latter would inevitably dominate the world. If social evolution was a branching affair, it was clear that one branch would have to be regarded as the main stem; the others, as mere side branches. Where Tylor's evolutionism resembles the anti-Darwinian theory of orthogenesis, Spencer's corresponds to Haeckel's pseudo-Darwinism, in which the principle of divergence is subverted by the need to treat European man as evolution's most advanced and most forward-looking product.

Spencer's belief that races would develop instincts related to their environment was based on the Lamarckian assumption, expounded in his *Principles of Psychology* as early as 1855, that long-established habits are eventually inherited as instincts. On the more general theme of progress, Spencer insisted that cultural and intellectual development go hand in hand. A higher level of culture stimulates greater use of the brain and thus increases the race's mental capacity, which in turn paves the way for further cultural growth (1870–71, 1:581–82; 1876–96, 1:23–29). Although Tylor had assumed the constancy of human nature, Spencer invoked Lamarckism to show how cultural development could only be achieved if the race also progressed intellectually. Whether progress was divergent or linear, the inherited effects of use and habit provided the most obvious way of relating cultural evolutionism to the more general evolutionary belief that the human race has advanced from an animal origin. As in biology, Lamarckism seemed a far more promising mechanism of progressive evolution than natural selection.

Historians of anthropology have recognized that Lamarckism provided the means of linking cultural evolutionism to the physical anthropologists' efforts to establish a hierarchy of racial types (Stocking 1968; Hatch 1973). A whole science of "craniometry" had been developed to measure the brain-sizes of the various races (Gould 1981). S. G. Morton in America and Paul Broca in France tried to demonstrate that the "lower" races occupied their inferior positions by virtue of their smaller cranial capacity and, hence, their lower intelligence. White anthropologists had frequently argued that the black races have certain apelike features (Haller 1971; Stepan 1982). An integrated theory of social, intellectual, and biological progress now seemed the most obvious way of explaining the hierarchy of races and cultures. Those races confined to a less-stimulating environment have experienced less pressure for intellectual development and hence have fallen behind in the advance along the physical, the mental, and the cultural hierarchies. "Lower" races certainly illustrated the Europeans' cultural ancestry, but they could also be seen as biological relics, trapped at a lower stage in the advance from the apes. Even Tylor, who would not adopt an evolutionary account of human origins, eventually conceded that some races have acquired a greater capacity for cultural development (1881, 74).

It would be wrong to claim that late-nineteenth-century physical anthropologists adopted a purely linear classification of the races. Superimposed on the idea of a hierarchy of mental ability was a wide recognition of the physical diversity of racial types. In theory, this was consistent with the view that evolution is both divergent and progressive, the white race being presumed to have advanced further than any of the other branches. There is little evidence, however, that the Darwinian theory of evolution had any major influence on physical anthropology. Some historians have argued that late-nineteenth-century anthropologists retained a pre-evolutionary view in which the races were seen as idealized types whose origins were of no real interest (Stocking 1968, chap. 3; Brace 1982). In fact, most theories of human origins which were current through the early twentieth century were designed to explain the supposedly vast antiquity of racial types. This was done by postulating that the division into racial branches had occurred *before* the acquisition of a fully human status (Bowler 1986). Several different lines of racial evolution were supposed to have independently advanced from an ape (or at least a prehuman form) up to a fully human level. This required a degree of parallelism that neither Darwin nor his modern followers could accept. Once again, the idea of multiple lines evolving in a predetermined direction had been applied to solve the problems of anthropology.

The belief that the cultural and mental progress of a race were interdependent had allowed a partial synthesis of cultural and physical anthropology. At the end of the century, however, this synthesis started to break down when the cultural anthropologists began to repudiate the link between the variety of cultures and the alleged hierarchy of racial characters. In Germany, Friedrich Ratzel (trans. 1896) refused to accept non-European cultures as a sign of racial inferiority and insisted that the culture of each region had developed in its own unique way. The British anthropologist W. H. R. Rivers also rejected the evolutionary hierarchy of cultures (1911; see Slobodin 1978; Langham 1981). Ratzel and Rivers both saw similarities between distant cultures as an indication of the diffusion of ideas from one area to another, which led them to abandon the idea of independent invention under the influence of a universal law of progress. In America a similar revolution in cultural anthropology was pioneered by Franz Boas and his followers, who insisted that social development was not constrained by biology and studied each society as a unique entity (Cravens 1978, chap. 2).

The popularity of linear progressionism had ensured that there would be no genuine Darwinian Revolution in nineteenth-century anthropology. Acceptance of mankind's animal ancestry was linked to the analogy with individual growth, not to Darwin's theory of geographically divergent evolution. At the level of pure analogy, both diffusionism and Boas's sense of cultural divergence as a historical process can be seen as compatible with a Darwinian model of branching, unpredictable evolution. But it would be meaningless to present the transformation of cultural anthropology in the twentieth century as a belated Darwinian revolution. The evolutionary hierarchy of races and cultures was abandoned, but Boas and his followers were so anxious to repudiate the link with race theory that they refused to admit any input from biology into the social sciences. Other movements in anthropology rejected a historical approach altogether and studied societies as though they were static entities, as in the "functionalism" of A. R. Radcliffe-Brown (Stocking 1984). It has been argued that the human sciences never adopted a truly Darwinian perspective (Greenwood 1984). Even the breakdown of the linear hierarchy was not accompanied by a transition from a typological to a populational view of human groups. Far from influencing the course of development

within anthropological theory, Darwinism found itself at first subverted by a rival model of evolution and then abandoned, along with
the whole idea that biology could serve as a foundation for the study
of culture. Efforts by the modern sociobiologists to reassert the importance of biology — now in a genuinely Darwinian form — have been
vigorously resisted by the social sciences (Caplan 1978).

Human Origins

The origin of the human species provided a focus at which biological and anthropological evolutionism intersected, and a study of
this topic will reveal the dominance of linear evolutionism. Traditionally, it has been assumed that the Darwinian Revolution in biology provided the impetus for a new evaluation of human origins. This
assumption is valid up to a point. Because of religious concerns,
Darwin and his followers knew that they would have to explain how
higher human faculties had emerged in the course of mankind's evolution from the apes. Tylor, on the other hand, could simply assume
that the earliest human beings already had the capacity for cultural
development. Yet there were other anthropologists and archaeologists
who thought that their interest in the primitive state of human culture
entitled them to hypothesize about the origins of mankind. John
Lubbock was a prominent Darwinist and Gabriel de Mortillet advocated an animal ancestry for mankind while most French biologists
still refused to accept evolution. Outside France, the rapid triumph
of evolutionism after the publication of the *Origin of Species* certainly
created the necessity for a reassessment of "man's place in nature."
But the anthropologists were already constructing their model of evolution, which was immediately available to anyone wishing to study
the emergence of mankind. If the anthropologists' linear progressionism coincided with at least one form of biological evolutionism,
this would provide an apparently plausible framework within which
to construct an account of how the human species had acquired its
highly developed mental and moral powers.

The Darwin industry has naturally tended to stress the contribution of the biologists. More surprisingly, it has defined a rather
limited arena within which it is prepared to explore the debate over
human origins. Even the more sophisticated historians who recognize
the oversimplification of the image of a "war" between science and
theology nevertheless tend to portray the reassessment of "man's
place in nature" as a two-sided negotiation between biology and religion, with psychology occasionally being admitted as a mediator.
The crucial question was whether or not the evolutionists could
justify the rejection of the traditional notion of the soul. Events that

followed the settlement of this debate in favor of the evolutionists are of little interest to the Darwin industry. Contemporary developments in anthropology and archaeology are seldom mentioned, presumably because they so clearly illustrate the role played by non-Darwinian concepts. For the same reason, the Darwin industry has also paid surprisingly little attention to the ongoing debates about *how* the human species might have evolved from the apes. A survey of post-Darwinian debates on human origins soon reveals that the issues raised by the Darwinian approach to the question — issues that are central to modern discussions — were seldom addressed in the late nineteenth century. Accounts of human origins invariably followed the developmental model, ignoring the kind of question that must be asked if evolution is seen as an open-ended process. My own recent study of this area (1986) was intended to reveal the extent of non-Darwinian thinking and was designed to encourage a dialogue between the Darwin industry and the historians of anthropology.

It is a striking fact that Huxley's *Man's Place in Nature* says virtually nothing about human origins. Although the book's title has been used to symbolize the debate by modern historians, Huxley contented himself with demonstrating the physical resemblances between humans and apes (Di Gregorio 1984). This was a valuable move, since it prevented conservatives like Owen from dismissing the whole idea of an ape ancestry for mankind. Yet Huxley refused to speculate about the crucial question of how human ancestors became differentiated from the apes, and he concluded with a comment about the antiquity of the human species that was more consistent with his own idea of the "permanence of type" than with Darwinism. There was thus a great deal left for Darwin to do if he wanted to provide a biological explanation of the origin of human faculties. The *Descent of Man*, published in 1871, was a powerful attempt not only to demonstrate the link with the animal kingdom but to explain how mankind acquired those faculties that were for so long regarded as uniquely human. It is a multifaceted book, deeply immersed in the progressionist assumptions of the time, but occasionally raising issues that were to be ignored for decades. To evaluate Darwin's role in the development of theories of human origins, it is first necessary to establish what the most popular view of the topic was in the 1870s and then to assess how this is related to Darwin's own account. The fragmented nature of historical research on these questions is illustrated by the fact that they remain controversial to this day.

The natural origin of human faculties had, of course, been proposed in Chambers's *Vestiges* of 1844. Although controversial when first published, Chambers's vision of an inherently progressive evolutionary trend began to seem increasingly attractive to those who found the trial-and-error aspect of natural selection distasteful. I have discussed how Chambers's essentially linear model of organic development was paralleled by the "evolutionism" of the cultural anthropologists. If *both* processes are seen as progressive, it would seem obvious that the development of human intelligence and culture are direct continuations of the process responsible for the ascent of life up to humanity's prehuman ancestors. There would be no need to postulate a sudden turning point in the evolution of mankind, only a continuous process of development. There might be some kind of threshold at which the Lamarckian interaction of intellectual and cultural growth would come into play, thus accelerating the already progressive tendency of animal evolution. If biological evolution is viewed as both divergent and progressive, it would then seem obvious that one branch might progress a little faster than the others and reach this threshold first, and this branch would give rise to the human race. This model raises mankind above the animals not because of some unique turn in human evolutionary history but merely because human beings have progressed further up a scale of mental development which all lines of evolution are potentially capable of ascending.

Such an account of human origins is thoroughly developmental and is compatible with the historicist character of nineteenth-century social evolutionism. But how does Darwin himself fit into this approach? Of those historians who have sought to define the essence of historicism, Mandelbaum (1971, 84) has portrayed Darwin as a typical exponent of this approach, while Nisbet (1969, 161) has argued that Darwin's vision of haphazard evolution was incompatible with the more popular developmentalism. Mandelbaum has even suggested that Darwin accepted the idea of a "main stem" of evolution leading to mankind. As already noted in chapter 2 above, many historians share Mandelbaum's conviction that progress played a vital role in Darwin's world view, although they may not endorse the implication that the key line of development was predetermined (see, e.g., Greene 1977, 1981; Young 1985a, 1985b). Like Spencer, Darwin saw progress and divergence going on simultaneously, in which case the "main line" would have to be defined merely as the branch that just happened to reach the threshold of cultural evolution first. Nevertheless, there are some accounts that have followed Nisbet in stressing the distinction between Darwin's theory of branching, adaptive evolution and Spencer's cosmic progressionism (Freeman 1974; Godfrey 1985). In this view, the essence of Darwinism is that each branch evolves in its own way, so that the "progress" of one branch cannot be measured by the standards of another.

Those who recognize a strong element of progressionism in Darwin's thinking, especially on human evolution, are concerned primarily with understanding Darwin's own motivations and the way he functioned within the social environment of his time. Although he gradually abandoned his belief in a benevolent Creator, Darwin was certainly inclined to hope that the white race did indeed represent the high point of an inevitable (if irregular) advance toward higher things. In contrast, the historians who stress the gulf between Darwin's biological theory and Spencer's progressionism are trying to pick out those aspects of the theory that transcended Victorian ideology. Whatever Darwin's own opinion, the lasting value of his work rests upon the nonprogressionist model of branching, adaptive evolution. From this perspective, Darwin's attempt to reconcile his theory with progressionism must count as a betrayal of his most important insights. He had pioneered an idea that had the potential to undermine the popular faith in progress, but he was unable to sustain its implications when discussing the more emotional topic of human evolution.

It seems rather silly to think of Darwin "betraying" the spirit of his own theory, but it is important to realize that his ideas could be understood in many different ways. Darwin himself was not a consistent progressionist; some of his purely biological discussions reveal an awareness of how difficult it would be to provide a progressionist interpretation of adaptive evolution. Nor were his contemporaries united in perceiving his theory as a contribution to progressionism. His opponents certainly did not see natural selection as an adequate mechanism of purposeful evolution. Some of Darwin's followers, including Huxley and the American Chauncey Wright, also tried to distinguish between the scientific theory of evolution and Spencer's progressionist cosmogony. If science was to be portrayed as a source of objective knowledge, it would have to dissociate itself from this kind of philosophical bias (Fichman 1984). By the end of the century, the nonprogressionist interpretation of Darwin's theory was reaching a wide audience through the science-fiction stories of H. G. Wells (Philmus and Hughes 1975). Under these circumstances, it would be misleading to single out the progressionist interpretation (even if favored by Darwin himself) and use this to portray "Darwinism" as a central component of nineteenth-century evolutionism. The theory was incorporated into mainstream progressionism only by stressing those aspects that had to be rejected in the formulation of modern biological Darwinism. If a label is required for the progressionist philosophy, "Spencerianism" would be better, since his thought was far more characteristic of the movement than Darwin's science.

The tension within Darwin's own thought can be discerned in the pages of the *Descent of Man*. Much of the book does indeed adopt a progressionist stance, implying that the white race and European culture represent the pinnacles of nature's advance. Darwin also made much greater use of Lamarckism in the field of human evolution, thereby moving closer to Spencer in this one area. Yet when he came down to the specific question of explaining how human ancestors were raised above the apes, he seems to have recognized the possibility of an answer that did not rely on simple progressionism. Despite his efforts to minimize the gulf between human and animal mentalities, Darwin knew that he must still account for the enormous advance in mental power from apes to humans. The progressionists tended to be so involved in reconstructing the sequence of development that they forgot to ask why the apes had been "left behind." They were unable to assume that the human branch has entered a phase of development not available to others, since this would have undermined the implication that progress toward the human level is inevitable. But in a theory of divergent evolution it is possible to imagine that the human race *is* unique after all, and to ask why its line of development has moved in a direction that differs so fundamentally from that taken by the apes. To the extent that Darwin was prepared to deal with the question in these terms, he was visualizing the appearance of the human species not as the necessary goal of progress but as the unpredictable result of a combination of circumstances that affected only mankind's ancestors.

Darwin's innovation occurs in chapter 4 of the *Descent of Man* (1871, 1:138–45) where he noted that mankind also possesses a unique *physical* characteristic, the upright posture. He speculated that human mental superiority over the apes may be a consequence of adopting a bipedal means of locomotion, which would have freed the human ancestors' hands for exploring the environment and, ultimately, for tool making. The use of the hand would have stimulated a more rapid growth of intelligence in this one line of evolution. The development of the human level of intelligence would thus not be predictable on the basis of a general progressive trend, since it could not have occurred without the crucial adaptation to a new means of locomotion.

Unfortunately, Darwin did not think very carefully about why man's ancestors adopted an upright posture. He hinted that it was precisely because it allowed them to use their hands more effectively, which fails to explain why the apes did not recognize the same advantages. It was A. R. Wallace who eventually filled in the missing element in Darwin's explanation, when he argued (1889, 459) that the change to an upright gait was an adaptation to a new environment.

Perhaps in response to some climatic change, man's ancestors were forced out of the trees onto the open plains, where bipedalism was a more appropriate means of locomotion. The apes stayed in the trees and retained their original level of intelligence. According to this theory, the expansion of the human brain was the indirect consequence of a change of habitat that was unique to humanity's ancestors.

Modern anthropologists suspect that the advantages of an upright posture in a plains environment may not be as obvious as Wallace supposed, but the fact remains that this is a very "Darwinian" way of looking at the question. The human mind is portrayed not as the inevitable outcome of a progressive trend but as the product of an unlikely combination of circumstances whereby an ape with the physical capacity for tool making was forced into an environment where that capacity was more actively exploited. Darwin and Wallace had constructed what is now called an *adaptive scenario* to explain the breakthrough that led to the emergence of mankind. Significantly, it is only within a theory of branching, adaptive evolution that it is necessary to specify such scenarios to explain the appearance of a particular form. The anthropologist Misia Landau (1984) pointed out that many later theories of human origins do indeed seem to have a storylike, or narrative, structure. This suggests an interesting parallel with traditional folk tales and creation-myths. But the narrative structure is also a very modern, that is, Darwinian, feature, since it represents an attempt to explain events that cannot be predicted simply on the basis of a pre-established trend.

The ideas of Darwin and Wallace may seem limited by modern standards, but they appear strikingly original when compared with the routine progressionism of their contemporaries. One searches the literature of the 1870s and 1880s in vain for any serious recognition of the need to specify a unique set of conditions that would explain why the human race has advanced so far beyond its closest relatives. The progressionists wanted to construct a sequence of development for the mental faculties, but they seldom asked why the human race alone had advanced to the higher levels of the scale. Ernst Haeckel supported Darwin's view that the upright posture was the first character that had distinguished man's ancestors from those of apes. But he made no effort to identify the benefits conferred by bipedalism and simply assumed that the subsequent development of speech was the real key to the emergence of the human mind (1876a, 2:299–300). A similar position was adopted by George John Romanes, who took on the task of defending Darwin's views of mental evolution against the critics who maintained an absolute distinction between the mental abilities of animals and humans. Romanes's *Mental Evolution in*

Man of 1888 accepted that language was the key to the development of rational thought, and speculated that the apes may have been blocked from this path by unsuitable vocal organs (154–55). Romanes failed to follow up this suggestion, however, and went on to assume that man's ancestors were already human in form, and already making tools, before they developed speech (1888, 370–76). He then fell back on the common belief that social life had helped to develop the human mind.

The link between social and mental development was expressed throughout the writings of Herbert Spencer. Yet neither Spencer nor Romanes thought it necessary to ask why the human race had advanced so much further along this path than any other animal species. Spencer's American disciple John Fiske attributed the more intense social life of man's ancestors to the fact that human children have to be protected through a longer infancy (1874, 2:342–44). Since the prolonged infancy was the result of increased intelligence, however, this merely threw the emphasis back onto a progressive trend toward mental development. Another American, the sociologist Lester Frank Ward, produced the most explicit statement of the underlying progressionism that blocked any recognition of the need to specify an adaptive scenario to explain human evolution. Ward argued openly against the idea that the human race possesses any unique physical characters lifting it above other animals (1895, 243–44). *Any* of the higher animals was potentially capable of reaching the threshold at which social and mental evolution began to reinforce one another. Whichever got to the threshold first would dominate the earth. Ward was evidently not interested in trying to explain why one line of progressive evolution had got slightly ahead of the others.

Darwin's suggestion that the adoption of an upright posture was directly responsible for tool making and hence for the dramatic increase in human intelligence remained largely unnoticed for over twenty years. Eventually, the controversy surrounding Eugene Dubois's discovery of Java man, *Pithecanthropus erectus*, in 1891–92 helped to focus attention on the priority of the upright posture (on human fossils, see Reader 1981). The *Pithecanthropus* remains consisted of a femur suggesting a bipedal means of locomotion and a skull fragment indicating a brain capacity between that of apes and human beings. Dubois himself was convinced that the ancestors of mankind had stood upright long before the final development of the brain. The Scots archaeologist Robert Munro now seized upon Darwin's point and argued at length for the idea that the use of the hand was crucial for the growth of human intelligence (Munro [1893] 1897). Wallace's suggestion that bipedalism was an adaptation to the new environment of the open plains could be linked to this line of argument. A number of early-twentieth-century theories of human origins at last began to focus attention on the possibility that a change of habitat played a role in separating human ancestors from those of the apes (Bowler 1986, chap. 7).

Paleoanthropologists remained committed to progressionism, however, and the "Darwinian" implications of Wallace's suggestion were easily subverted. The anatomist Grafton Elliot Smith defended the view that a steady expansion of the brain throughout primate evolution offered an explanation for the emergence of mankind (Smith [1912] 1924). Although not a Lamarckian, Smith held that the arboreal habits of the early primates had stimulated the development of active hands and brains. The immediate human ancestors had not been driven out of the trees by a climatic change; they had moved out onto the plains deliberately once their intelligence had risen to the level where they could appreciate the even greater possibilities of this new environment. In this way the importance of the adaptive scenario was minimized. Far from being a crucial transition, the move onto the plains merely continued a long-established trend toward increased intelligence. As revealed by the phylogenetic tree he constructed for the primates, Smith still saw a "main line" of development aimed at the production of mankind (see fig. 7). The tree's side branches represented those primates that had sacrificed their flexible bodily structure through overspecialization for life in the original arboreal habitat.

Smith's main line of human evolution is reminiscent of Haeckel's efforts to discern a central theme running through the animal kingdom toward mankind. Significantly, it was a disciple of Haeckel, Gustav Schwalbe, who most strongly defended a linear arrangement of the known hominid fossils as an illustration of the pattern of human evolution (1904, 1906, 1909). In Schwalbe's view, Dubois's *Pithecanthropus* and the now-familiar Neanderthal type of early humanity represented stages in the continuous ascent from the apes to the modern human form. This neat linear pattern soon began to break down, however. In France, Marcellin Boule seized upon a newly discovered Neanderthal specimen to argue that this apelike race was too primitive to be the ancestor of modern human beings (1909–11, 1921). Boule was motivated by a desire to discredit the linear evolutionism of de Mortillet, who had linked progressionism to the hope of future social reform (Hammond 1982). Boule now argued that the human evolutionary tree consists of many branches and that the Neanderthals were a distinct species of mankind which had not advanced as rapidly as the ancestors of modern humans. In Britain, a similar view was championed by Arthur Keith (1915), who linked it with the dis-

Figure 7. Phylogenetic tree of the primates, from Elliot Smith (1924).

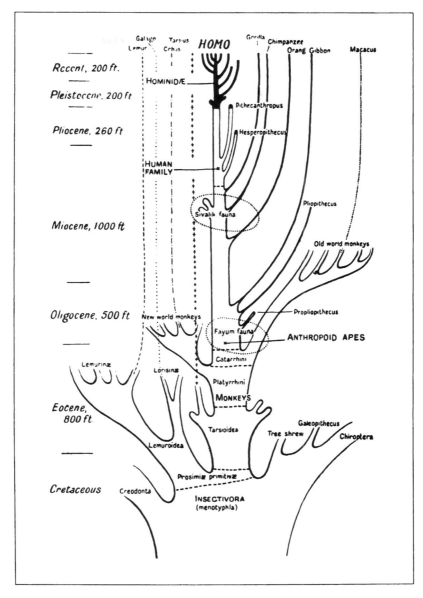

covery of "Piltdown man" in 1912, a find exposed as a hoax in 1953 (Weiner 1955).

At first sight, this transition to a multiple-branched model of human evolution may seem like a belated acceptance of Darwinism, but in fact it was a continuation of the old progressionism in a new and more sophisticated guise. The new theory had the merit of being compatible with the physical anthropologists' belief that the human "species" consists of a number of distinct racial types. Boule and Keith now saw human evolution as a process in which many branches advanced at different rates toward the goal of a fully human form. In other words, they explicitly evoked the image of parallel evolution characteristic of anti-Darwinian paleontologists such as the American neo-Lamarckians and the supporters of orthogenesis. Arthur Smith Woodward, best known for his discovery of the Piltdown remains, was a leading advocate of orthogenesis who appealed openly to this theory as an explanation of human origins (1915, 3). Orthogenesis also featured strongly in W. E. Le Gros Clark's first study of human ancestry (1934, 288). Nor were these isolated instances; in the decades before the emergence of the modern synthetic theory of evolution, the majority of paleoanthropologists, like the paleontologists, favored theories heavily laced with parallelism and the concept of directed variation (Bowler 1986, chap. 8).

Although superficially more sophisticated than the linear progressionism of Haeckel and Schwalbe, early-twentieth-century theories of human origins retained the progressionist and almost teleological character established by the anthropologists and non-Darwinian biologists of the previous century. Only after the emergence of the synthetic theory in the 1940s did anything like a Darwinian interpretation of human evolution begin to take shape. Le Gros Clark (1955, 1967) now abandoned his support for orthogenesis and became a leading advocate of the theory that bipedalism represents a fundamental breakthrough in establishing the hominid line of evolution, best illustrated by the Australopithecine fossils of South Africa. If animal evolution was to be explained solely in terms of adaptation, it became necessary at last to specify an adaptive scenario for the origins of mankind, instead of relying on a predetermined trend. The current preoccupation with the attempt to understand *why* human ancestors became bipedal and eventually acquired larger brains is thus a relatively modern development within a field that retained a largely non-Darwinian character well into the twentieth century.

Was there, then, a "Darwinian Revolution" in the study of human origins, and, if so, when did it occur? The debate over "man's place in nature" may have intensified after the publication of Darwin's theory, but it was resolved by the establishment of an evolutionary, rather than a truly Darwinian, account of how the higher faculties might have developed. Darwin himself made an attempt to create a theory in which the human race was not portrayed as the inevitable goal of progressive evolution, but there is little evidence that he was able to disturb his contemporaries' faith in teleology. Anthropologists, archaeologists, and non-Darwinian biologists combined to establish a progressionist view in which the production of the human mind remained the central purpose of creation. If the social sciences of the early twentieth century turned their backs on progressionism, they did so along with Western culture in general, and for reasons that go far beyond any developments in biological theory. The scientific study of human origins remained insulated from this loss of faith in progress, and it did so by retaining an explicitly non-Darwinian view of the evolutionary process. If there was a Darwinian Revolution in this field, it occurred only in the 1940s, when the modern synthesis in evolution theory finally made it impossible to believe that nature had been striving throughout evolution to create the human form.

Chapter Seven

SOCIAL DARWINISM

✲ The debate launched by the *Origin of Species* had an impact on the emergence of a progressionist view of human history, but it seems clear that biology was but one factor among many that contributed to the growing optimism of nineteenth-century thought. Victorian society saw itself as the pinnacle of human achievement and justified its opinion by picturing the whole of history as a steady progression toward an intended goal. Even when this sense of confidence faltered, as it did in the *fin de siècle* atmosphere of the 1890s, the threat of cultural degeneration was often invoked only as a warning that the conditions necessary for progress needed to be re-established (Nordau 1895; see Chamberlin and Gilman 1985). But what *were* the conditions most likely to encourage progress? Did the process of biological evolution provide a model for the establishment of social policies? The assumption that biological evolution is progressive led naturally to the claim that human progress could best be maintained by following the lessons taught by nature. Social evolutionism thus took progress for granted and sought inspiration from biology on the best way to ensure that the trend would continue.

In the classic image presented by Hofstadter ([1944] 1955), the predominant way of linking biological and social progressionism was through *social Darwinism*. Struggle was the essential means of progress in the evolution of life, and hence struggle was both inevitable and beneficial in human society. Hofstadter presented Herbert Spencer's extremist interpretation of laissez-faire individualism as the vehicle that transmitted an essentially Darwinian view of society from Britain to America. Since Darwin accepted Spencer's phrase "survival of the fittest" as a legitimate expression of how natural selection works, there must have been an underlying congruence between the Darwinian emphasis on struggle and the ruthless attitude of nineteenth-century capitalism. Darwin merely projected onto nature the values of a society obsessed with "progress through struggle," and his theory thus became a valuable weapon in the armory of social thinkers seeking to legitimize this ideology by claiming that struggle was an integral part of nature.

I have already found it necessary to challenge this interpretation of the link between Spencer and Darwin. In order to establish the originality of Darwin's thinking (chap. 2 above), I argued that the theory of natural selection drew upon utilitarian individualism in a way that differed significantly from Spencer's philosophy. To substantiate the claim that Darwin's theory, or at least the aspect of it taken most seriously by modern biologists, was not welcomed by his contemporaries, I must now challenge the image of social Darwinism as a valid characterization of late-nineteenth-century thought. In part, the challenge will draw upon the work of historians who have questioned the extent to which Darwinian metaphors were, in fact, used by the apologists of capitalism. But even more central to my argument is a demonstration of the enormous diversity of ways in which the images of struggle and selection could be applied to society. Given this variety, I shall argue that it is meaningless to depict natural selection as the "most obvious" scientific expression of what was in reality an extremely diffuse enthusiasm for the idea of struggle.

The real problem is to define what is meant by "social Darwinism" (Himmelfarb 1968; Halliday 1971; Rogers 1972; Bannister 1979; Bellomy 1984; La Vergata 1985). Even those historians who stress the movement's importance recognize several different ways in which Darwin's theory can be translated into a social policy. In addition to the link with laissez-faire capitalism, Hofstadter distinguished two lesser forms of social Darwinism. In one, the level of competition is switched from the individual to the group, providing a justification for the struggle between nations and races in which the weaker are conquered or exterminated by their superiors. In the other, the emphasis is placed on artificial rather than natural selection; the supporters of eugenics advocated a selective breeding policy to improve the biological character of the race, especially by discouraging unfit individuals from breeding. Each of these alternatives drew upon the Darwinian theory in a different way, resulting in a series of social implications that are, in fact, incompatible with one another. Their only common denominator is that they have all come to be regarded as evils by later generations of liberal and socialist thinkers. "Social Darwinism" has become an epithet used to castigate almost any attempt to include biology as a determinant of human behavior.

Even if the link between Darwinism and capitalism turns out to be less secure than Hofstadter imagined, one could argue that the extension of the theory's influence into the areas of nationalism and eugenics confirms the dominant status of Darwinian imagery in late-nineteenth-century thought. Whoever the targets and whatever the means of elimination, these policies all exhibit a ruthless indifference toward those who have failed in the drive toward progress. But the variety of the applications, coupled with the fact that eugenics did not become popular until the early years of the twentieth century, throws doubts on any simple interpretation of the relationship between Darwinism and social thought. The metaphors of struggle and selection were so malleable that they could be applied in many different ways and could be retained by a later generation of writers who had abandoned individualism. Even if Darwin's own thoughts had been inspired by a particular interpretation of utilitarian individualism, the images created by his theory could be exploited equally well by ideologues with diverse interests. Darwinian language was certainly used by writers who saw success as the reward conferred by nature upon those who contribute toward progress, but those writers were seldom inspired by — and often did not understand — the principle of natural selection. Far from shaping the thoughts of a whole generation, Darwin's theory was merely a tool that could be adapted to a wide range of purposes.

This point becomes even more obvious when we realize that Darwin's name was also invoked by social thinkers who were opposed to the ruthless attitudes outlined above. The term *reform Darwinism* (Goldman 1952) identifies the precepts of a group of writers who used Darwinism to justify their call for a social policy based on cooperation, as a means of winning mankind's struggle against the physical environment. Greta Jones's (1980) study of British social Darwinism has revealed the frequent use of the theory by liberals and others opposed to the competitive ethos of individualism. In both France and Germany, laissez-faire social Darwinism was ignored, and social analogies tended to link the theory with reform or with national competition (Kelly 1981; Clark, *Social Darwinism in France*, 1984). Late-nineteenth-century social thought was a good deal more complex than the simple image of social Darwinism implies. The exponents of success at any cost did not have it all their own way, and the move toward a more centralized society steadily gained momentum even in the English-speaking world. The fact that all varieties of opinion made appeals to Darwinism indicates not that natural selection was the leitmotif of the time but that the theory has no special affinity with any particular social policy. The appeals were inspired by the desire to create a spurious air of scientific authority rather than by the theory itself.

Reform Darwinism shows that it is possible to associate the theory with a viewpoint quite alien to that of the more familiar social Darwinism. Indeed, the reform Darwinists' call for increased social cohesion resembles the policy that Hofstadter (1955, chap. 4) associated with an *opponent* of Darwinism, the Lamarckian sociologist Lester Frank Ward. Many Lamarckians believed that a unified society should be able to direct its evolution along a chosen path. The term *social Lamarckism* has not gained the same degree of acceptance as its Darwinian equivalent, but it is clear that the rival theories of biological evolution could also be used as the basis for social analogies. Social evolutionism goes far beyond social Darwinism, and if Darwin's theory had only limited influence on science, it is logical to assume that the biological alternatives were exploited in the outside world. The popularity of the non-Darwinian forms of social evolutionism may simply have been obscured by the historians' tendency to focus on Darwinism. Recent research has not only uncovered the prevalence of non-Darwinian analogies but has also suggested that the social policies once associated with Darwinism were often supported by appeals to non-Darwinian theories. Thus the eugenics movement drew its supporters from the ranks of Darwinians, Mendelians, and Lamarckians.

At this point the claim that a scientific theory can express the values of a particular ideology degenerates into absurdity. If it turns out that virtually *any* theory can be used to justify *any* social policy, merely by tinkering around with the way in which the analogy is drawn, then it becomes impossible to believe that a theory can ride to popularity on the coattails of a particular ideology of which it is the most "obvious" scientific expression. Biological theories are often associated with social policies and may well be evaluated in terms of what their social implications are perceived to be. But the creation and popularization of these associations is itself a social process, not an indication that each theory has an intrinsic link to a particular ideology. The claim that nineteenth-century thought created an environment within which Darwinian principles could flourish therefore collapses, not just on the evidence that the role of Darwinian metaphors has been exaggerated, but on the impossibility of defining a cultural environment that would automatically have generated support for this as opposed to any other theory.

The Struggle for Existence

In his reassessment of the role played by social Darwinism in American thought, Bannister (1979) argued that the specter of rampant capitalism justifying itself by the motto "survival of the fittest" is a myth created by the opponents of laissez-faire individualism. Although the robber barons of American industry may have occasionally used Darwinian metaphors to vindicate their commercial dominance, there were very few self-advertised social Darwinists. "Social Darwinist" was almost always used pejoratively by reformers as a way of highlighting the ruthlessness of their opponents. It suited the reformers of the late nineteenth century to pretend that Darwinian attitudes had encouraged the competitive ethos of capitalism. Mid-twentieth-century liberals, including many historians, preferred to support rather than challenge this myth, because it was in their own interests to emphasize the evils of the past that were now being eliminated. Bannister held that the myth can be demolished by noting the extremely limited use of Darwinian language even by those capitalists who are supposed to have been the chief advocates of individualism. All too often, "social Darwinism" turns out to be Spencerianism, based on Herbert Spencer's quite different interpretation of the benefits of laissez-faire (Freeman 1974).

A more recent survey by Donald C. Bellomy (1984) confirmed that the Spencerians did not call themselves "social Darwinists." The term was first used in the debates of the European sociologists in the 1880s and only became common in America after 1900. Yet leftist historians have strongly resisted the tendency to separate Darwin from social Darwinism (Young 1985b; Moore 1986a). They have argued that Darwinism was integrated into the social debates from the very beginning, and they see Bannister's revisionist historiography as a right-wing plot to establish the "objectivity" of the science used to uphold biological determination. It is true that some historians of science, disturbed by the unpleasant connotations of "social Darwinism," have claimed that Darwin did not associate himself with the harsh social values of his time. This certainly goes too far: Darwin *did* share these values (Greene 1977) and may well have been partially inspired by them in the creation of his theory. If he dismissed some races or social classes as inferior or unfit, it is hardly surprising that his work was seized upon in an attempt to uphold these prejudices. Darwinism *was* social—but does this justify the use of "social Darwinism" as a label for the whole range of determinist ideologies which has flourished since the mid-nineteenth century?

The socialist historians' refusal to give up their use of "social

Darwinism" suggests that they have inherited the mantle of the earlier liberals who coined the expression. They want to use Darwin's name because, unlike Spencer, he still commands attention as a scientific innovator. If Darwin can be linked to determinism, he will provide a neat symbol of the ongoing attempt to use science as a foundation for right-wing ideologies. But if the most popular vindication of individualism was based on Spencer's Lamarckism, it is surely misleading to use Darwin's name as a label instead of Spencer's. The social historian may find the distinction trivial, but the historian of science must recognize that Darwin's most innovative and (ultimately) successful ideas were at first subordinated to Spencer's quite different interpretation of the struggle metaphor. It is certainly necessary to investigate how the theory was used in this way, but to treat Darwin as the founder of social Darwinism is to promote the false impression that natural selection formed the chief model upon which ruthless social policies were based. This in turn promotes the belief that selectionism was swept to popularity because it so accurately reflected the values of its time. Darwin's status as an original thinker is reduced by making him just another Spencerian. Then insult is added to injury by demoting Spencer from his status as a Victorian prophet and applying Darwin's name to the whole system. The Spencerian aspects of Darwin's thought should not be concealed, but neither should the differences between the ways in which the two men explored biological evolutionism. The struggle metaphor could be exploited in a number of different ways; Spencer is forgotten and Darwin remembered precisely because there was something in Darwin's theory that could be used by later generations of biologists raised in a very different scientific and cultural environment.

The confusion arises partly because Darwin and Spencer shared the then-popular assumption that some races are inherently inferior to others. In this respect they can indeed be regarded as pioneers of biological determinism. But the concept of a racial hierarchy was so popular that many biological theories, including Lamarckism, were used to justify it. When applied to the white race, however, Spencer's philosophy was *not* rigidly deterministic, since it was based on the belief that most individuals can improve themselves when provided with the appropriate stimulus. The great paradox of nineteenth-century social thought was that lower races were condemned to permanent inferiority, while the white race was supposed to be capable of continuous progress. This paradox was particularly acute for Lamarckism and may account for that theory's gradual replacement by a more consistent determinism. But Spencer did not rely on the element of determinism inherent in the theory of natural selection to justify his claim that free enterprise is the best guarantee of social progress. Instead, he maintained that free enterprise provides the best equivalent of the conditions under which natural progressive evolution takes place. The biological component of his thinking came not from an assumption that the unfit are condemned by heredity but from a direct comparison with the process of struggle that was supposed to be the driving force of natural evolution. Spencer's philosophy and Darwin's theory both invoked the "struggle for existence" and can thus be seen as parallel manifestations of a system in which the "fit" demand the right to exploit those less active than themselves. But, while parallel, these concepts were not equivalent, since the struggle metaphor could be exploited in two different ways.

Chapter 2 outlined the essential differences between Spencer's early advocacy of laissez-faire and the Darwinian concept of natural selection. Spencer was already thinking in evolutionary terms by the time he wrote his *Social Statics* (1851) to emphasize the importance of individualism for social progress (Peel 1971; Kennedy 1978). Even at this stage, he accepted that the unfit must suffer and may even be eliminated by starvation. It should thus come as no surprise that he later coined the expression "survival of the fittest" to capture the essence of the Darwinian mechanism. Yet Spencer did not independently discover the theory of natural selection, because he always saw the elimination of the unfit as a purely secondary process. The real purpose of the suffering experienced by society's failures is not to kill them but to teach them in the most effective way possible that they must seek a new and more productive way of life. Spencer hoped that the majority of the unfit would thus be forced to become fitter. Spurred on by harsh necessity they would learn to participate more fully in society and would teach their children the importance of thrift and industry. The Lamarckian implications of this last point were not lost on Spencer, and this is why he consistently supported Lamarckism as the most important mechanism of social and biological evolution throughout the rest of his career. Like many Lamarckians, Spencer failed to distinguish clearly between biological inheritance and the transmission of new habits by learning. But his failure in this respect merely demonstrates how difficult it was for the nineteenth-century mind to comprehend distinctions that the Mendelian Revolution has imposed upon modern thought.

Spencer's philosophy was thus a form of social Lamarckism, since it assumed that the population would learn new habits adapted to the changing social environment. Unlike many later Lamarckians, however, Spencer did not believe that the human race could direct its own future evolution by controlling the habits imposed upon new generations through the educational system. He assumed that social evolution is so complex a process that the human mind cannot predict what will be needed to speed things up. Government interference in the behavior of individuals will almost certainly backfire. Nature thus has to be left to take its own slow but certain course, even if this means a great deal of suffering. Only in the distant future might there be a perfect balance between individual behavior and society. Spencer's distrust of socialism was based on his belief that ameliorating the lot of the poor will remove the one stimulus that might teach them to adopt more productive habits. In later works such as *The Man versus the State* (1884), Spencer's opposition to the steady drift toward interventionist policies took on an even more strident tone that threw greater emphasis on the elimination of the unfit. He now accepted that there might be a greater proportion of individuals who are incapable of benefiting from nature's lessons. Yet his continued support for Lamarckism suggests that he still saw the elimination of the unfit as a secondary mechanism, ultimately less important than the stimulation of personal development in those who respond positively under pressure.

The American businessmen who took Spencer as their philosophical mentor were looking for a means of justifying their own economic power. His Synthetic Philosophy promised them that worldly success is nature's reward to those who lead the way in adapting to the progress of industrial society. Misery is the punishment imposed on those whose improvidence blocks their participation in the same trend — a punishment that can always be escaped by doing better in the future. No doubt some tycoons occasionally resorted to Darwinian language, but it was often in a context that would have puzzled Darwin. In a passage frequently quoted by historians (e.g., Hofstadter 1955, 45), John D. Rockefeller declared:

> The growth of a large business is merely a survival of the fittest. . . . The American Beauty rose can be produced in the splendor and fragrance which brings cheer to its beholder only by sacrificing the early buds which grow up around it. This is not an evil tendency in business. It is merely the working-out of a law of nature and a law of God.

As Rockefeller implied in his last sentence, this was not exactly a new message inspired by the theory of natural selection. On the contrary, Spencer's philosophy offered at best a modernized version of liberal Protestant theology, an endorsement of traditional values by nature as well as by God which included the promise of mankind's perfection in this world rather than the next (Moore 1985a). Many of the businessmen were ostentatiously religious, and there was no shortage of clergymen such as Henry Ward Beecher willing to tell them that evolution represented the unfolding of God's will.

Nor were the appeals to the "survival of the fittest" as frequent as Hofstadter has suggested (Wilson 1967; Russett 1976, chap. 4). Once allowance is made for the fact that much Spencerian language can sound rather Darwinian to those not alert to the differences between the two approaches, it becomes difficult to identify more than a handful of explicit appeals to Darwin's theory. The Yale sociologist William Graham Sumner — a classic social Darwinist by Hofstadter's standards — stressed only the struggle of mankind against nature and refused to adopt a Darwinian interpretation of struggle *within* society (Bannister 1979, chap. 5; Garson and Maidment 1981). Sumner was even criticized by the Spencerians for doubting that the free-enterprise system would inevitably lead to progress. Among those who did follow Spencer, John Fiske, in his *Outlines of Cosmic Philosophy* (1874), used an explicitly religious tone based on the assumption that progress entails more cooperation and social harmony. The claim that businessmen such as Andrew Carnegie were really social Darwinists has also been disputed. Carnegie had a healthy respect for democracy, and in his "Gospel of Wealth" proclaimed himself the steward of the nation's riches, responsible for using them for the good of society (Wall 1970; Bannister 1979, chap. 4). The rags-to-riches philosophy of many small businessmen drew upon pre-Darwinian values deeply rooted in American society. In addition, the small businessman was unlikely to look kindly upon a system that would encourage his bigger competitors to drive him to the wall (Wyllie 1954, 1959).

In Britain, Spencer's philosophy reflected the commercial and professional classes' dislike of the aristocracy, and Darwinism was also seized upon by liberals seeking an excuse for challenging traditional authority (Jones 1980, chap. 3). Walter Bagehot's *Physics and Politics* (1872) sought to apply "natural selection" to society by arguing that free enterprise is the only way to encourage the emergence of cultural variations that can be tested by experience. The liberals certainly wished to see a more open society, but they were convinced that each generation produces its own natural aristocracy. The important thing was to allow nature's mental and moral elite to gain control of society. All too often, this meant only a continuation of the old hierarchical system under new management. The extreme individualism of Spencer's philosophy might appeal to businessmen, but the professional class was now beginning to look upon itself as the natural aristocracy and wished to retain a stable social order.

Evolutionism was called in merely to argue that the top position should be occupied by those who thought they most deserved it.

Jones (1980) emphasized the sheer variety of ways in which images derived from Darwin's theory were applied in British social thought. Far from being linked to a single "social Darwinism" based on struggle, Darwinian images were in fact employed by a wide range of political philosophies. Even the socialists could argue that Darwin was on their side, encouraged perhaps by the example of Alfred Russel Wallace's claim that equality is the only way of ensuring that the fittest men and women will marry one another (Durant 1979). The socialist J. Keir Hardie drew upon the notion of "group selection" found in Darwin's *Descent of Man*:

> Darwin stated emphatically that "those communities which included the greatest number of the most *sympathetic* members would flourish best", and in so stating he conceded the whole case for which the socialist is contending. It is sympathetic association and not individualistic competition which makes for progress and the improvement of the race. (Hardie 1907, 94)

This was a valid point, but it utilized only one facet of Darwin's overall theory. The call for a more cooperative society was increasingly associated with those who opposed the theory of natural selection. Such calls did not come from the socialists alone. The émigré Russian anarchist Peter Kropotkin published a series of articles in the 1890s, subsequently collected under the title *Mutual Aid* (1902). Here he argued that evolution itself works not through struggle but through enhancing the cooperative instinct. A more cooperative human society is thus merely the natural end product of the evolutionary trend. Significantly, Kropotkin later wrote in support of biological Lamarckism (1912). The claim that the drive toward altruistic behavior illustrates a divine purpose underlying the evolutionary trend toward mankind was also expounded by the Scottish cleric Henry Drummond in his aptly titled *Ascent of Man* (1894).

Far from dovetailing neatly into a single kind of social philosophy, Darwinism was clearly drawn in on both sides of a complex ideological debate. The supporters of individualism and of cooperation drew freely on both Darwinian and non-Darwinian metaphors whenever it suited them. The role played by evolutionary ideas in the move toward a more centralized society will be explored below, but it is necessary, first, to note the use of Darwinism to justify competition at the national and racial levels. Hardie's appeal to Darwin's notion of group solidarity was unfortunate in the sense that the same concept could all too easily be used in support of an ideology that most socialists found repugnant, namely, imperialism. If the focus of natural selection in human evolution was switched from individual to group competition, then emphasis could be placed on the cooperation within the group that would develop to ensure the group's success against its rivals. Modern nations could be seen as the end products of centuries of group selection, still locked in combat with each other. The British paleoanthropologist Arthur Keith eventually created a whole theory of human evolution based on this idea (1948). In the late nineteenth century, some social thinkers had already begun to realize that the Darwinian struggle for existence might take place between rival national or racial groups. Bagehot's *Physics and Politics* (1872) portrayed human history as a series of cultural experiments, with successful new cultures dominating their neighbors and forcing them to adopt the new way of life. By the last decade of the century laissez-faire individualism was in retreat, but the power struggle between rival Western nations was increasingly being seen as a struggle for existence in which the strongest would eventually dominate the world (Hofstadter 1955, chap. 9; Bannister 1979, chap. 12).

In France and Germany, nationalism was the most prominent means through which the struggle for existence was applied to human affairs (Clark, *Social Darwinism in France*, 1984; Zmarzlik 1972). It has been argued that Ernst Haeckel's Monist League, founded to support the materialism expressed in his *Riddle of the Universe* (trans. 1900), was an important vehicle for transmitting social Darwinism to the early members of the Nazi party (Gasman 1971). Haeckel, however, never adopted natural selection as an explanation of how new species were formed; his "Darwinism" always concentrated on the struggle between rival species or subspecies. Other historians have therefore challenged the assumption that the conventional form of social Darwinism played an important role in the writings of Haeckel and his contemporaries (Kelly 1981). But even in the English-speaking world, there was a temporary wave of enthusiasm in the 1890s for the work of Friedrich Nietzsche, which could easily be interpreted as favoring a policy of "might is right" (Bannister 1979, chap. 10). Yet Nietzsche had ridiculed Darwinism, confirming that his enthusiasm for strength of personality took its inspiration from quite different sources. If Darwinism played any role in creating the Nazis' half-baked mythology of struggle, it did so along with a host of other, often incompatible, strands of thought.

The Nazis were able to draw upon a long tradition of European thinking on the race question (Poliakov 1970; Tennenbaum 1956). At the turn of the century, the Western powers were engaged in a bitter struggle to annex the "uncivilized" parts of the globe. This presupposed that Europeans have the right to dominate the rest of the human race because of their mental and moral superiority. Chapter 6 above described how the anthropologists had established a hierarchy of racial types long before Darwin's theory emerged and how the essentially non-Darwinian developmental viewpoint remained the most influential factor supporting efforts to rank the various races on an evolutionary scale of perfection. The black and brown races were all too often portrayed as the occupants of lower places on the scale. Although many Darwinians adopted the conventional attitude toward race, it can hardly be said that Darwin's particular theory of evolution flourished because it could be used to support the racial hierarchy. In fact, the image of branching, haphazard evolution was difficult to reconcile with the belief that the white race is in some absolute sense the most developed form of the human species.

If Darwinism influenced the debate on race, it did so by encouraging the belief that the races must be competing with one another and that the white race is destined to come out on top. As imperial expansion took over from laissez-faire as the dominant ideology, Darwinian language was invoked to justify the conquest and even the extermination of the "unfit" races. The leader of the biometrical school of Darwinism in Britain, Karl Pearson, penned the following passage:

> It is a false view of human solidarity, a weak humanitarianism, not a true humanism, which regrets that a capable and stalwart race of white men should replace a dark-skinned tribe which can neither utilise its land for the full benefit of mankind, nor contribute its quota to the common stock of human knowledge. The struggle of civilised man against uncivilised man and against nature produces a certain partial "solidarity of humanity" which involves a prohibition against any individual community wasting the resources of mankind. (Pearson [1892] 1900, 369)

In a footnote here Pearson warned against the brutality of simple extermination but still rejoiced that the natives of America and Australia were being replaced by civilized whites. Pearson certainly knew his Darwinism and evidently saw nothing wrong with extending the idea of survival of the fittest into the area of race relations. In these circumstances, other less knowledgeable writers can hardly be blamed for assuming that this was a legitimate application of the Darwinian metaphor.

The idea of race conflict was used by early-twentieth-century paleoanthropologists to explain key events in human evolution. In his

Ancient Hunters of 1911, W. J. Sollas argued that the human species had not evolved at a uniform rate over the whole globe. Occasional episodes of rapid, local evolution had created new and higher racial types that had then spread out to exterminate their inferior neighbors. This idea was taken up by Arthur Keith and others to explain the apparently sudden disappearance of the Neanderthal race from Europe. Keith drew an explicit parallel with the modern world: "Those who observe the fate of the aboriginal races of America and Australia will have no difficulty in accounting for the disappearance of *Homo neanderthalensis*. A more virile form extinguished him" (Keith 1915, 136). Sollas expounded the ethical implications of this viewpoint quite explicitly.

> What part is to be assigned to justice in the government of human affairs? So far as the facts are clear they teach in no equivocal terms that there is no right which is not founded on might. Justice belongs to the strong, and has been meted out to each race according to its strength. . . . Hence it is a duty which every race owes to itself, and to the human family as well, to cultivate by every possible means its own strength: directly it falls behind in the regard it pays to this duty, whether in art or science, in breeding or in organization for self-defence, it incurs a penalty which Natural Selection, the stern but beneficent tyrant of the organic world, will assuredly exact, and that speedily, to the full. (Sollas 1911, 383)

At first sight this quotation looks like a classic expression of social Darwinism, suggesting that Darwin's theory was still capable of inspiring new scientific and ideological applications fifty years after its original publication. Yet in contrast with Pearson, Sollas and his fellow paleoanthropologists were certainly *not* Darwinians in the sense that they accepted natural selection as the mechanism of progressive evolution (Bowler 1986). Later in the same book, Sollas dismissed natural selection as an "idol of the Victorian era" and claimed that it was incapable of creating new characters in evolution (1911, 405). This apparent expansion of Darwinism into a new area came at a time when the credibility of the selection theory among biologists was at an all-time low. Something is surely wrong with the whole concept of "social Darwinism" if it is based on the use of Darwinian language by biologists or political writers who do not accept the Darwinian theory in its most important area of application.

How was it possible for an avowed opponent of biological Darwinism to write about the human races in what appears to be Darwinian language? To answer this question it is only necessary to recall Spencer's quite different application of the struggle metaphor to human affairs. Spencer coined the expression "survival of the fittest" to denote what was to him a purely secondary mechanism of evolution: natural selection merely eliminates the less successful products of the primary and essentially Lamarckian process by which struggle stimulates the acquisition of new habits. Sollas and Keith were not Lamarckians, but they did have a largely developmental view of evolution in which the production of new and higher types was seen as inevitable. Struggle was again viewed merely as a means of eliminating those races that fall behind in this purposeful advance toward higher forms of humanity. Both the individualist and the racial applications of the struggle metaphor ignored the basic principles of Darwin's theory by subordinating natural selection to an almost teleological mode of development. Throughout its chequered career, social Darwinism was thus Darwinian in name only; it employed Darwinian metaphors at a superficial level and ignored the theoretical principles that were to inspire the revival of interest in Darwinism by the biologists of the 1930s and 1940s.

Obviously there were many adherents of the "might is right" philosophy throughout the late nineteenth and early twentieth centuries, although there was a switch of the favored application from laissez-faire individualism to imperialism. Despite their common appeal to the struggle metaphor, these two ideologies were in fact hostile to one another. Spencer railed against the growth of imperialism in the later part of his life, and the imperialists (as will be discussed below) rejected laissez-faire in favor of social management. The rival social policies could appeal to the theory of natural selection because both were using Darwinian language at an essentially superficial level. Even if we believe that Darwin himself was influenced by laissez-faire individualism, it is not possible to assume that capitalism provided a cultural environment in which natural selection would inevitably flourish at the expense of all other evolutionary theories. The robber barons of American industry would have welcomed Spencer's philosophy whether or not Darwin's theory had been available to inspire the convenient phrase "survival of the fittest." The imperialists treated the extermination of "inferior" races as inevitable even though many of them did not accept the selection theory as it had been proposed by Darwin. What modern biologists perceive as the core of Darwin's theory represented an application of the struggle metaphor which was significantly at variance with the once-popular assumption that the main thrust of biological and social evolution, although harsh in some of its applications, is directed by a more purposeful force.

Biological Management

The quotation from Kier Hardie above shows how a particular interpretation of Darwinism could be used to support socialism rather than unrestrained free enterprise. The whole concept of reform Darwinism rested on the assumption that Darwin's writings could somehow be used to vindicate the "struggle for the life of others" (Goldman 1952). If this seems a rather strained interpretation, it must be admitted that many calls for social reform were, like Kropotkin's, based on a non-Darwinian vision of evolution. Even so, Benjamin Kidd's widely read *Social Evolution* (1894) managed to argue for limited reforms as a means of preserving the competitive stimulus of the free enterprise system (Crook 1984). Kidd's point was that unrestrained competition can easily leave the poor so depressed that they will cease to have any interest in society. Reforms that give them a fighting chance will encourage the process of social evolution. Kidd noted that irrational factors such as religion had played a similar role as a social lubricant in earlier periods.

As an anarchist, Kropotkin hoped to build a society based on spontaneous cooperation, without centralized state control. Most Lamarckians agreed that evolution would enhance the cooperative instincts, but in general they preferred to see the state as an important force that should direct social development. Lamarckism played a significant role in the late-nineteenth-century social sciences, generally on the side of reform (Stocking 1962, 1968). Perhaps the leading figures in America were Lester Frank Ward and Joseph LeConte (Hofstadter 1955, chap. 4; Scott 1976; Stephens 1982). Lamarckism was important because it allowed social reform to be presented as a means of improving the moral character of the human race. Accepting that natural evolution is progressive, the reformers challenged Spencer's assumption that mankind cannot hope to control the process. Instead they argued that human beings were now able to take charge of their own evolution, accelerating the rate of progress and directing its course along a deliberately chosen path. Their view was that social reform will improve people's mental and physical health, and the improvements will be inherited by the next generation and will thus have permanent benefit for the race. Furthermore, by using education to shape the development of young minds, new instincts can be produced that will permanently improve the moral character of mankind. As late as the 1920s, the biologist at the center of the "case of the midwife toad" could still generate newspaper headlines by proclaiming that Lamarckism allows the production of an improved form of humanity (Kammerer 1924; see Koestler 1971).

Although Spencer had used Lamarckism in support of laissez-faire, the reformers turned the theory into an almost utopian vision of mankind's immediate future. By the end of the century the need for the state to exercise more control was widely accepted, but other forces were undermining faith in the Lamarckians' promises. The imperialists had their own very different reasons for advocating a strong centralized state (Semmel 1960). National security demanded a unified state to maintain the population at a high level of efficiency. Karl Pearson called himself a socialist (1894), but his real concern was the need for central control to meet the threat posed by rival powers. Spurred on by British failures in the early stages of the Boer War, Pearson (1901) argued that a more efficient society was essential if the empire was to survive. In America, the Progressive era ushered in a wave of support for state-controlled capitalism under which the whole nation was to be managed as efficiently as individual companies.

It could be argued that the transition to a harsher policy of social management represented an acceptance of Darwinian principles ignored in earlier decades. The Lamarckians, whether supporters or opponents of Spencer, had assumed that evolution is progressive and that individual personalities can be improved in a way that is significant for the future of the race as a whole. They merely disagreed over whether mankind can direct the natural progress toward a higher state. But as T. H. Huxley pointed out in his 1893 lecture on "Evolution and Ethics," the whole point of Darwinism is that natural evolution is *not* progressive (Huxley 1893–94, vol. 9; Helfand 1977; Paradis 1978). Whatever his earlier failure to appreciate the message underlying Darwin's theory, Huxley's philosophy of "cosmic pessimism" had now led him to believe that human ethical values are incompatible with the principles of a harsh and ultimately meaningless nature. Spencer was wrong because leaving everything to nature will not guarantee progress. Faith in the inevitability of progress did not vanish overnight, but in the following decades many social thinkers took an increasingly cautious tone. Progress might still be possible, but it was not inevitable, and society would have to be actively managed to ensure that it developed in the right direction.

Perhaps the clearest expression of this more authoritarian and less optimistic philosophy was the emergence of the eugenics movement. The term *eugenics* was coined by Darwin's cousin Francis Galton in 1883 to denote the application of a controlled breeding policy to the human race (Forrest 1974; Buss 1976; Cowan 1972b, 1977). In his *Hereditary Genius* of 1869, Galton had already argued that the son tends to inherit his father's abilities, and he had suggested that it is important for the future of the human race that individuals with high abilities should have large families. As time went on, he became even more concerned about the apparently high rate of increase among the poorer social classes, which he assumed consist of individuals with the lowest levels of ability. A eugenics policy would promote breeding among the fit, while discouraging or even preventing the unfit from reproducing their defective characters. At first Galton was largely ignored — a fact which confirms that the Victorians did not think of their society in Darwinian terms. Hardly anyone in the 1870s or 1880s could tolerate a philosophy that so rigidly limited the individual's ability to respond when challenged by adversity. Only at the turn of the century did Galton suddenly find himself at the head of a rapidly growing movement to present eugenics as the most effective way of managing the biological future of the race (Blacker 1952; Allen 1975b; Bajema 1977; Farrall 1979; Kevles 1985; on eugenics in America see Haller 1963; Pickens 1968; Ludmerer 1972; in Britain see Mackenzie 1982; Searle 1976, 1979; Jones 1986).

The eugenics movement enjoyed spectacular growth in the first decade of the twentieth century. In Britain, Galton founded the National Eugenics Laboratory in 1904 and, soon afterward, the Eugenics Education Society. The American Breeders Association set up a Eugenics Committee in 1904, and the Eugenics Records Office was founded in 1910 to coordinate research. The first International Eugenics Congress was held in 1912. All were agreed on the need to prevent the feebleminded and insane from reproducing, either by institutionalization or sterilization. But the social pressures behind this wave of activity differed from country to country. The Eugenics Records Office was funded by the Rockefeller and Carnegie foundations as an example of what could be achieved by planned capitalism (Allen 1986b). Soon the movement was playing an active role in the campaign to limit the immigration of "unfit" races into the United States. In Britain, the professional classes were most deeply involved, seeing eugenics as a way of avoiding the tax burden that would be required to finance the elimination of poverty. A strong eugenics movement flourished in Germany long before the Nazis came to power, based on the claim that the health of the population should be scientifically managed (Weiss 1986). The common factors uniting these forces were the belief that the state should actively control the biological fitness of its population and the assumption that direct interference with human breeding is the only sure way of achieving this goal.

Hofstadter (1955, chap. 8) presented eugenics as a form of social Darwinism based on the idea that natural selection must be replaced with its artificial equivalent in a civilized society. Galton himself certainly drew an analogy with artificial selection. His leading British follower was Karl Pearson, who combined scientific Darwinism with a keen awareness of the struggle for existence among nations. Pearson saw eugenic control of the population as vital for national survival. Mackenzie (1982) argued that Pearson's statistical work was designed to ensure that biometrical Darwinism lent support to eugenics. It is also possible that R. A. Fisher's pioneering work in population genetics was inspired by his commitment to eugenics (Box 1978; Bennett 1983). At a more profound level, the fact that eugenics sought to control the population in violation of natural trends seems to reflect a less optimistic view of evolution more in tune with what are now taken to be the most important principles of Darwinism.

Yet the association between eugenics and Darwinism should not be exaggerated. Once again, what looks like a belated emergence of Darwinian values turns out to be explicable in terms that are more general than a simple link to the theory of natural selection. Galton himself did not believe that natural selection could explain the origin of new characters and appealed instead to saltation directed by an internal force (1889, 18–34). Pearson's support for the selection theory was genuine enough but was clearly not representative of biological opinion at the turn of the century. The surge of enthusiasm for eugenics came at a time when the majority of biologists had turned their backs on Darwinism. Even when natural selection began to attract more attention in the 1930s, there was no direct link to eugenics. If R. A. Fisher supported eugenics, the co-founder of population genetics, J. B. S. Haldane, wrote openly against the movement from a socialist perspective (1938). In America, the most effective scientific support for eugenics came from early Mendelians such as C. B. Davenport. As noted in chapter 5 above, at this point the Mendelians were hostile to Darwinism and did not believe that natural selection played any role in the evolution of new characters. The popularity of eugenics must thus be accounted for in terms of broader social factors. It was certainly not the result of Darwinism promoting the idea of artificial selection of the human population.

How was it possible for eugenics to be upheld by appeals to both Darwinian and anti-Darwinian theories? The movement did not draw its inspiration from Huxley's anti-progressionist view of natural evolution, which had been intended to support a quite different kind of interventionist policy. Its basic values were shaped by the rise of what today would be called biological determinism. In the emerging debate over whether human nature is shaped primarily by nature (inheritance) or nurture (education and upbringing), the earlier generation of Lamarckians were all on the side of nurture (Pastore 1949; Cravens 1978). They believed that the individual can be improved by his or her upbringing in a way that will permanently benefit the race. Eugenics was a manifestation of growing support for the rival "hereditarian" position, based on the assumption that the individual's essence is fixed by biological inheritance and cannot be improved by education. Its supporters held that improving conditions for the "unfit" poor is a waste of money; the only way of maintaining the race's biological fitness is by preventing these unfortunates from breeding. Such a position could certainly be justified by analogy with artificial or natural selection, since these are both fundamentally hereditarian mechanisms. But Darwinism is just one among many possible applications of the hereditarian view of character-determination. The early geneticists gained their influence in science by demonstrating that characters are determined by Mendelian factors. They could thus explain why it was important to prevent the spread of unfit characters, even though they attributed natural evolution to the spontaneous appearance of favorable mutations without the need for selection. Genetics and Darwinism could thus provide equally plausible vindications of eugenics, and in America, especially, it was genetics, with its claim to be the first "scientific" study of heredity, that was seen as the more appropriate foundation.

Darwin had not convinced his contemporaries that the individual's fitness depends on the luck of its inheritance rather than its own efforts. This parallels his failure to convince any more than a handful of followers that evolution does not progress inexorably toward a morally significant goal. These two points only began to gain wide acceptance in the early twentieth century, as faith in progress began to wane and as the Mendelian Revolution created the first effective scientific foundation for hereditarian principles. Once it had been demonstrated that the majority of genetical mutations are harmful, it became easier to accept that poor inheritance creates a permanent barrier impeding the individual's self-improvement. This in turn seemed to confirm that nature offers no guarantee of progress to the species. Biologists and social thinkers now began to see that the "progress" of a population must depend upon selection by an external force. Once the logic of this situation was accepted, it was inevitable that biologists would eventually realize that adaptation offers the only selective force capable of directing natural evolution. But the re-emergence of Darwinism as a force in modern biology should not mislead us into thinking that the selection theory played a central role in the creation of the new value system, particularly in America. It was the Mendelian Revolution, not the Darwinian, which was most closely associated with the emergence of the new values.

It is thus at best only a half-truth that eugenics was a belated recognition of Darwinian principles; rather, the emergence of hereditarian attitudes in society encouraged scientists to modify a number of competing theories to give apparent support to the new ideology. Darwinism was one such theory, but even Lamarckism could be adapted to the trend by assuming that the inheritance of acquired characters takes place so slowly that racial characters are fixed in the short term. Lamarckians had always accommodated themselves to the concept of a racial hierarchy, but early-twentieth-century defenders of the theory such as E. W. MacBride used this point actively to support eugenics (Bowler 1984b). In the English-speaking world, though, it was Mendelism that most clearly represented the scientific foundation of hereditarian policies, if only because genetics had become the most visible symbol of scientific management. Even when the majority of geneticists turned their backs on the growing excessiveness of the eugenicists' demands, popular writers continued to exploit the assumption that genetical factors are the chief determinant of character.

The modern sociology of science recognizes that theoretical developments are shaped by the social and professional environment (Mulkay 1979; Barnes and Shapin 1979; Shapin 1982). But the sociologists also accept that it is impossible to draw simple one-to-one relationships between scientific theories and particular social values. A certain cultural environment may stimulate the development of an appropriate view of nature, but it will often be possible for several different scientific theories to play the same role. It may appear to a historian that one theory was more easily adapted to the social environment than its rivals—but can the historian be sure that this impression has not been created by hindsight? The establishing of links between theories and values is itself a social process. Success may depend on the particular circumstances of time and place, not on any intrinsic rapport between a theory and the values with which it has been most popularly linked. The claim that Darwinism was the most obvious way of projecting late-nineteenth- or early-twentieth-century values onto nature fails because it does not recognize the role played by rival theories in satisfying these values, and because it is inconceivable that the same theory could have remained the clearest expression of values that changed significantly through time. Darwin's theory was exploited by a wide range of political thinkers, many of whom did not appreciate the principles that would subsequently lead to its acceptance by modern biologists. There were thus many different forms of social Darwinism, of which the only ones to be remembered under that name today are those cited by modern social thinkers as examples of the attitudes they reject.

Eugenics was based on the assumption that the individual's intelligence is determined by biological inheritance. But determinism could take other forms, with quite different social implications. Writers such as Kropotkin had, in effect, argued that human behavior is conditioned by social instincts that have been shaped by generations of evolution. In the early twentieth century some social psychologists developed theories based on the claim that human beings possess a range of instinctive behavior-patterns (e.g., MacDougall 1908). Evolution certainly played a role in these theories, although Lamarckism and group selection seemed equally plausible explanations of how such instincts could be formed. Non-Darwinian evolutionary ideas continued to shape the thought of some influential early-twentieth-century sociologists, including Vilfredo Pareto (Nye 1986). In general, however, the social sciences now began to emancipate themselves from biology (Cravens 1978). Anthropologists such as Franz Boas argued that human behavior is shaped almost entirely by cultural factors, not by biologically imprinted instincts. They also repudiated the claim that ability is predetermined by inheritance. The social sciences have remained hostile both to determinism and to any effort to argue that biological and cultural evolution are shaped by similar processes.

Any subsequent attempt to reintroduce biology as a determinant of human behavior has been met by sociologists who raise the specter of a renewed social Darwinism. In some cases, it is by no means clear that Darwinism has actually played a critical role in shaping the theories to which the sociologists object. The so-called anthropology of aggression of Konrad Lorenz (1966) and Robert Ardrey (1966, 1976) might seem to rest upon a Darwinian view in which human nature is shaped by combative instincts. But in Ardrey's case, the creation of these instincts is attributed to the hunting life style of early human ancestors, not to natural selection's role as the mechanism of evolution. Conversely, many contributors to the Modern Synthesis of Darwinism and genetics have adopted an optimistic view of mankind's position in the evolutionary process (Huxley 1957; Dobzhansky 1962; Simpson 1967; see Greene 1981). While recognizing that evolution is not directed toward a single goal, they have nevertheless stressed its "creativity" and have insisted that it upholds a traditional view of ethics. These Darwinists have been only too anxious to throw off the old image of social Darwinism.

The re-emergence of biological determinism in recent years has been prompted to a large extent by the success of molecular biology

(Kaye 1986). Success in cracking the genetical code, not an interest in natural selection, has encouraged this new breed of biologists to insist that heredity plays a major role in human affairs. Only in the case of sociobiology is there a genuine link between evolutionism and the human sciences (Wilson 1975, 1978; Caplan 1978). Here modern Darwinism has made a genuine effort to extend its range by explaining apparently altruistic instincts in terms of natural selection acting at the individual rather than the group level. If it were accepted that such instincts do at least partially control human behavior, then the determinist thesis would be renewed in a new and more vigorous form. Naturally enough, those who resist this move have found it convenient to use the term *social Darwinism* as a means of identifying sociobiology as the heir to a long tradition of biological determinism. Yet, as Michael Ruse (1986) pointed out, if sociobiology can become the basis for a genuine social Darwinism, the process will have to be regarded as the first real attempt to take Darwinism (as opposed to evolutionism) seriously in human affairs.

Chapter Eight

A CULTURAL REVOLUTION?

✸ The Darwinian Revolution has a significance outside the history of science because the advent of evolutionism is seen as a watershed separating modern culture from the traditional roots of Western thought. This interpretation of the revolution is shaped by the broader image of the warfare between science and religion. The aptness of this image is the one point of agreement between those who feel in tune with the revolution's consequences and the critics who see it as an unmitigated disaster. The warfare metaphor was created by writers such as Draper (1874) and White (1896) who saw science as an important weapon to be used in their struggle to destroy the superstitions that had long kept the human spirit in chains (Moore 1979). Darwin's highly visible destruction of the biblical creation-myth symbolized the breaking of these chains. Rational thinkers then felt free to establish a new and more realistic view of human nature. According to this model, the Darwinian Revolution is an episode in the spiritual liberation of mankind.

Liberal humanism survives, but its optimism does not reflect the growing anxieties of the modern world. The most extreme form of reaction against the direction taken by modern culture finds the warfare metaphor equally useful. Creationism, whether in the Tennessee of the 1920s or in modern Louisiana and Arkansas, sees Darwin as a symbol of the forces that have destroyed the traditions from which Western culture derived its moral fiber (Durant 1985b). The creationist view is that Darwin's alternative to divine creation undermined faith in both revelation and the afterlife and thus helped to set up a society in which sensual gratification is the only goal. There are many other critics who, although not advocating a return to biblical literalism, nevertheless distrust the materialist element in modern thought and see Darwinism as a major source of that element.

The previous chapters have already exposed the weakness of this confrontational model as an interpretation of evolutionism's historical impact. For every conservative theologian who rejected Darwin, there were several liberals who welcomed the idea that natural development was the unfolding of a divine plan. Many ostensibly religious thinkers were quite willing to accept Spencer's claim that material success was a measure of the individual's contribution to social progress. Social Darwinism was a modernized version of the Protestant ethic and owed its harshness as much to the basic character of that ethic as to the means by which it was modernized. Some historians now regard the advent of evolutionism as a byproduct of the ideological revolution through which the professional and commercial classes wrested authority away from its traditional sources, the church and the landed aristocracy. Certainly, the evolution debate was but one among many factors responsible for the secularization of European thought (Chadwick 1975). Seen from this perspective, the traumatic aspects of the *Origin* debate were necessarily short-lived. Whatever the distaste at first engendered by the possibility that human beings were descended from apes, the demand for a progressionist world view was strong enough to force most thinkers into a fairly rapid acceptance of the basic idea of evolution.

Progress was the key to the new ideology, and thus it was essential for evolution to be seen as a process with a moral purpose. Even if we choose to ignore the sociological explanation of the Darwinian Revolution, the fact remains that most late-nineteenth-century evolutionism retained a progressionist or developmental component that would not be accepted by modern biologists. Students of nineteenth-century culture accept that many applications of evolutionary principles did not employ what is now regarded as the essence of Darwin's thinking. Morse Peckham (1959) has suggested the use of "Darwinisticism" to denote the cultural equivalent of what I have labeled "pseudo-Darwinism" in biology. I prefer "pseudo-Darwinism," partly because it is less clumsy, but also because it more clearly implies that the principles of modern Darwinism are not involved. The important point, however, is to recognize the extent to which the evolutionism of the immediate post-Darwinian era was shaped by non-Darwinian values. Since those values included the retention of teleological principles inherited from an earlier period of Western thought, it is necessary to query the extent to which the *Origin* debate can be seen as the focus for a dramatic transition to materialism. The scientific revolution precipitated by Darwin's catalytic effect on biology necessitated a fairly abrupt confrontation between developmentalism and certain traditional ideas about mankind's role in nature. But the fact that so much of the old value system could be accommodated within the developmental version of evolution ensured that the traumas would soon be forgotten. The Darwinian Revolution was completed so quickly because it did not require acceptance of the more radical aspects of Darwin's thought.

Both the supporters and the opponents of modern Darwinism accept that the theory's most important feature is its elimination of any need to see change as goal-directed. The ambivalence of Darwin's own views on progress and teleology reveals the tension generated by his partial realization of the fact that his theory had the potential to destroy the value system within which it had been conceived. The belated recognition of Darwinism's radical implications by modern biologists suggests that the developmental view of evolution was abandoned somewhere in the transition from nineteenth- to twentieth-century values. The real cultural revolution occurred not with the *Origin* debate but with the twentieth century's loss of faith in the progressionism that had allowed so many traditional values to be retained. Historians who dislike Darwinism for its moral implications (e.g., Barzun [1941] 1958; Himmelfarb 1959) stress its importance because they see it as a symbol of the materialism that has had such a catastrophic effect in the modern world. The Darwinian Revolution was crucial not for its immediate effects but because it paved the way for a later, far more comprehensive rejection of traditional values.

Acceptance of mankind's animal ancestry certainly helped to shape the ideas of many of the thinkers who directed the twentieth century toward a determinist or irrationalist view of human nature. But what role did Darwin's conception of the evolutionary process play in the fundamental revolution marking the transition to modern thought? The nonteleological character of modern scientific Darwinism might encourage a belief that the theory of natural selection helped to undermine the general faith in progress. Yet everything we have seen so far suggests that the formative years of twentieth-century thought were dominated by non-Darwinian views of evolution. It was a developmental model of social change that allowed Marx to proclaim the historical inevitability of revolution and enabled Freud to infer the prehuman character of much subconscious mental activity. We should not ignore the role played by *evolutionism* in the creation of many facets of the twentieth century's fragmented culture. But there is little to suggest that a renewed interest in Darwin's theory played any more than a minor role in the emergence of a pessimistic world view. The triumph of Darwinism in biology has not been accompanied by any broad acceptance of Darwinian principles in human affairs; indeed, the social and human sciences have remained hostile to all biological models.

Evolution and the Crisis of Faith

The evolution debate certainly played a role in what has often been called the "Victorian crisis of faith." In the English-speaking world, at least, the intellectual life of the early nineteenth century was still governed to a significant extent by concepts derived from the Christian religion. As the century progressed, secular values that had already begun to gain ground in continental Europe began to filter into Britain and America. At midcentury there was a dramatic shift of opinion in which many traditional interpretations of Christianity were abandoned or reconstituted along more liberal lines. The rise of evolutionism prompted a revaluation not only of mankind's relationship to the rest of the cosmos but also of the way in which the moral and physical worlds were supposed to be governed by their Creator. A few daring but as yet untypical thinkers rejected the notion of a Creator altogether. Among these traumatic events, the *Origin of Species* was but one among many sources of anxiety for the religious community (Symondson 1970). The publication of *Essays and Reviews* (Jowett et al. 1860), a collection of articles urging a more liberal view of Scripture and of God's relationship to nature, provoked an even more violent controversy (Willey 1956; Brock and MacLeod 1976).

Even within science, Darwin's book was not the only source of opposition to traditional values. The debate over Chambers's *Vestiges* had already made plain some of the implications that would follow from evolutionism. Although scientists had united with conservative theologians in condemning the book, the resulting publicity ensured that many ordinary people would read it (Millhauser 1959). By this time many radical thinkers among the lower classes had already begun to accept evolutionism as a necessary consequence of rejecting traditional values (Desmond 1987). The possibility that the mind might be a product of the physical activity of the brain (not of a spiritual entity, the soul) was implicit in the teachings of phrenology. Although often dismissed as pseudo-scientists, phrenologists such as Franz Joseph Gall tried to show that mental faculties were located in certain parts of the brain (Young 1970; Cooter 1985). By provoking a relatively sudden conversion of the scientific world to evolutionism, the *Origin* merely forced intellectuals to face up to consequences implicit in a much wider trend toward a naturalistic philosophy. In the post-Darwinian world, the physicist John Tyndall's notorious "Belfast Address" of 1874 provided an alternative source of arguments for materialism.

The more vocal exponents of the new "scientific" philosophy certainly liked to picture themselves as being at war with theology. The warfare metaphor (with the assumption that science was winning) occurs frequently in the writings of T. H. Huxley, for instance. Yet Huxley was not intrinsically hostile to a religious view of the human situation; his campaign was against *theology*, not against religion itself (Barton 1983). Most Victorians, including a majority of the scientists themselves, retained the view that the material universe must have a moral purpose, and they were determined to interpret science in a way that would not destroy that sense of purpose (Turner 1974). Theology was a threat because it represented a traditional source of intellectual authority that had always kept science in a subordinate position. If the scientific profession was to gain the status it thought it deserved within the new industrial world, it would have to establish its role as the most authoritative source of rational knowledge. Evolution was useful to professionals such as Huxley precisely because it allowed them to demonstrate that theology was obliged to give ground in an important area of human understanding (Turner 1978; Fichman 1984).

The emergence of science as a recognized professional activity was itself a consequence of broader movements within society brought about by the Industrial Revolution. The commercial and professional classes were increasingly aware of their importance within an industrialized economy and were anxious to gain political power at the expense of the landed aristocracy. In the interpretation offered by Robert M. Young (1973, 1985a) and James R. Moore (1982b, 1986b), the emergence of evolutionism was part of the ideological struggle to establish this new social order. The middle classes needed a world view in which progress was inevitable in order to legitimize their challenge to traditional authority. The transition to a society in which success depended on ability, not on aristocratic origins, would be justified if it could be portrayed as a step in the moral progress of the universe. Yet the very fact that a moral purpose was still assumed meant that at the intellectual level this was a revolution *within* the tradition of natural theology, not *against* that tradition.

As the leading philosopher of evolutionism, Herbert Spencer did not seek to destroy the hierarchical structure of society, nor did he wish to disparage the old Protestant virtues of thrift, industry, and sobriety. But he did want to argue that those who exercised these virtues were entitled to rise toward a higher social position. He justified this demand by claiming that their efforts would lead to overall social evolution. Instead of a static society offering rewards in the next world, evolutionism created a dynamic system with rewards in this world, and it claimed that these rewards were the inevitable consequences of natural laws. The new morality required not the destruction but the reinterpretation of traditional values, with nature rather than God as their most immediate guarantor. For many less adventurous thinkers, however, it was easier to retain the belief that God was the ultimate originator of the natural system that rewarded effort and punished sloth. If biological evolutionism was to become the scientific foundation of this new system, it would have to offer apparent support for the claim that individual effort contributed toward the moral progress of the race.

The problem was that Spencer's abstract arguments for Lamarckism were not enough to convince biologists that evolutionism worked in science. The publication of the *Origin of Species* was thus a godsend to those who saw Spencer's approach as the key to a new ideology. By making evolutionism respectable in science, Darwin provided the exponents of social evolutionism with a valuable propaganda weapon — whether or not they understood the biogeographical arguments that underpinned the new theory. The comparatively rapid acceptance of evolutionism during the 1860s revealed by Ellegård's (1958) survey of the British periodical press can be explained as a consequence of the theory's congruence with the new ideology. According to this model, Darwin succeeded not through the strength of his scientific arguments but because of social pressures demanding a conceptual foundation for the new view of human affairs. As in science itself, the pressure to burst the restrictions of the static world view had been building up for decades, and the *Origin* merely provided the stimulus that prompted the necessary rethinking of traditional ideas. The bitterest opposition to the theory naturally came from conservative political and religious thinkers who saw creationism as the basis for a static model of society which would preserve aristocratic privileges. The sense of a traumatic conversion to evolutionism was experienced most acutely by the less adventurous liberal thinkers who balked at the rejection of certain Christian images required by the new progressionism.

The most troublesome of these conventional images was the assumption that mankind alone was blessed with a spiritual essence, most clearly revealed by the capacity to make moral judgments. Conservative thinkers were appalled at the prospect of an animal ancestry for the human soul. Disraeli's famous response to the question "Is man an ape or an angel?" typifies this reaction: "My lord, I am on the side of the angels" (Monypenny and Buckle 1929, 2:108). When Huxley debated Wilberforce at the 1860 British Association meeting, it was the remarks on human ancestry which prompted the biggest outcry. Historians now have a less positive view of Huxley's achievement on that day (Lucas 1979), and it must be remembered that many scientists shared the distaste for evolutionism's tendency to link mankind with the apes. Fears about human ancestry had been at the forefront of the opposition to Chambers's *Vestiges*. Charles Lyell found this a stumbling block to complete acceptance of evolutionism even in the post-Darwinian era (Bartholomew 1973). Wallace, too, was driven to postulate a supernatural agency that controls the evolution of the human mind (Smith 1972; Kottler 1974; Turner 1974). The evidence suggests, however, that many thinkers soon overcame this problem. After 1870 there was a steady decline in opposition to the basic idea of an evolutionary origin for mankind, in part because the progressionist model of evolution assumed that nature's activities had a moral purpose. The most important human characteristics were thus transferred *into* nature, thereby eliminating the need to preserve an absolute distinction between the material and spiritual worlds.

From the start it was clear to some religious thinkers that the conflict with evolutionism could be resolved by paying more attention to a line of argument which had, in fact, been pioneered by Chambers's *Vestiges*. Transmutation need not usher in a wave of materialism if God's creative power could be seen as the driving force of all natural developments. As Charles Kingsley wrote to a fellow clergyman in 1863,

> Darwin is conquering everywhere, and rushing in like a flood, by the mere force of truth and fact. The one or two who hold out are forced to try all sorts of subterfuges as to fact, or else by invoking the *odium theologicum* [sic]. . . . But they find that now they have got rid of an interfering God — a master magician as I call it — they have to choose between the absolute empire of accident, and a living, immanent, ever-working God. (Kingsley [1876] 1877, 2:171)

Kingsley explained in his "Natural Theology of the Future" of 1871 (Kingsley 1890) that he had no doubt that science promoted the latter view of development. Although controversial at first, the idea that God works through law rather than capricious miracles soon became the basis for a general reconciliation. Far from being in conflict with science, the new understanding of religion would allow both sources of knowledge to work hand in hand. More radical thinkers might follow Spencer in refusing to speculate about the origins of a morally purposeful universe, but the majority were willing to accept evolution as the unfolding of a divine plan. Religious writers such as Henry Drummond insisted that the moral as well as the physical world was governed by law, providing an easy compromise for anyone wishing to reconcile social evolutionism with traditional Protestant values (Moore 1982b, 1985b; Livingstone 1987).

It is within this context that we should evaluate Gertrude Himmelfarb's controversial claim that Darwin merely precipitated a "conservative revolution" (1959, chap. 20). She argued that the *Origin* "was the catalyst that broke the crust of conventional opinion. It expressed and characterized what many had obscurely felt" (452). The shock was one of recognition, not of discovery, and "in practical morals and beliefs the post-Darwinian world was not so different from the pre-Darwinian one" (411). These opinions anticipate those of modern radical historians but leave a question about how much of Darwin's creative scientific insight was incorporated into the superficially transformed world view. Himmelfarb has often been criticized for her unsympathetic portrait of Darwin and his theory, and it is clear that the full-scale materialism she has associated with the selection theory did not become a major component of late-nineteenth-century thought. "Natural selection" may have been on everyone's lips, but few understood the concept. The young Henry Adams, for instance, recalled his own uncritical acceptance of evolutionism as typical of many intellectuals. Writing of himself in the third person, he described his reactions thus:

> He was a Darwinist before the letter; a predestined follower of the tide; but he was hardly trained to follow Darwin's evidence. . . . Darwin hunted for vestiges of Natural Selection, and Adams followed him, although he cared nothing about Selection, unless for the indirect amusement of upsetting curates. He felt, like nine men in ten, an instinctive belief in Evolution, but he felt no more concern in Natural than in unnatural Selection. (Adams 1918, 225)

Ellegård's (1958) survey of the periodical press also shows that the rapid conversion to evolutionism was not accompanied by any general enthusiasm for natural selection. The revolution was conservative in character, not because everyone had already accepted materialism but because the most radical interpretation of evolution — Darwinian natural selection — was seldom anything more than a stimulus to the emergence of rival, non-Darwinian views of organic and social development.

The need to reconcile science and religion made design a question of paramount importance. If evolution was the unfolding of a divine plan, how did the Creator ensure that His ends would be achieved? The new ideology promoted by liberal thinkers was based on the claim that individual actions are the ultimate source of racial progress. What individuals do determines both their own material rewards and the course of society's evolution. Spencer's emphasis on effort and achievement was central to this message. But natural selection left individuals at the mercy not only of a harsh environment but also biological processes over which they had no control. Conservative theologians and scientists were quick to point out that the selection of random variations would usher in what Kingsley called the "absolute empire of accident." As a result, even liberal thinkers tended to shy away from a specifically Darwinian interpretation of the struggle metaphor. Paradoxically, as Moore (1979, chap. 11) pointed out, a few of the staunchest Calvinists were able to accept the purposeless suffering inflicted by natural selection as compatible with their view of human sinfulness. The American botanist Asa Gray tried to argue that whatever the mechanism of evolution, the process could still be seen as a reflection of the Creator's purpose (essays 1876; see Dupree 1959). But even Gray conceded in the end that variation must be purposeful rather than random, thereby coming into line with the popular preference for a more obviously teleological model of evolution. Although Gray himself did not favor Lamarckism, many of the religious thinkers who were disturbed by the inhumanity of natural selection eventually followed Spencer in opting for use and effort as the factors controlling inherited variation.

Conservative thinkers were reluctant to see progress as the result of individual effort, but they were not locked into a rigid creationism. They had no objection to the divine plan of nature being interpreted in a developmental fashion. The modern tendency to explain evolutionism as a byproduct of bourgeois values can all too easily lead us to forget that idealism made a comparatively rapid adjustment to the new state of affairs. Sensing which way the wind was blowing, idealist naturalists such as Richard Owen now made a valiant effort to show that evolution could be understood as a pattern imposed by the Creator's mind upon nature, designed to unfold whatever actions might be taken by individual organisms. By exploiting the analogy between growth and evolution, it was possible to argue that the whole plan of development was somehow predesigned, just as Chambers had proposed in *Vestiges*. The resulting "theistic evolutionism" flourished in the 1870s as the most vocal source of opposition to Darwin's theory (Bowler 1977a, 1983, chap. 3). In their determination to show that evolution represents the unfolding of a divine plan, many naturalists argued that there are lines of evolutionary development which cannot be explained in terms of how individual organisms react to their environment. The duke of Argyll's *Reign of Law* (1867) and St. George Jackson Mivart's *Genesis of Species* (1871) were perhaps the most effective vehicles promoting this approach. Kingsley (1890, 313) noted that Mivart's views exactly fitted his requirement for a system in which the scientist could actually see the divine hand at work. Mivart's book was widely hailed as a telling critique of Darwin's theory. It raised a host of objections against natural selection and presented evidence to show that evolution has progressed under the influence of some inbuilt guiding force.

At one level, Mivart's attempt to see evolution as the unfolding of a divine plan was short-lived. By the 1880s it was becoming unfashionable for scientists to make explicit appeals to the divine will as an explanatory factor. To this extent, the evolutionary naturalism of Darwin and Spencer had triumphed. Yet Mivart's arguments were extended in a modified form by the late-nineteenth-century exponents of orthogenesis. They, too, believed that evolution is projected along a predetermined course over which the individual organisms have no control, but they invoked variation-pressures arising within the germ plasm to explain the trends. The popularity of this alternative to Darwinism in certain areas of biology confirms that there were some naturalists who wanted to preserve a bulwark not only against natural selection but also against the whole idea that development is shaped by the interaction between organisms and the environment. As the initiative slipped away from Spencerian individualism, concepts of internally programmed evolution flourished for a while, although the ultimate beneficiaries would be the nondevelopmental hereditarian theories of the early twentieth century. The combination of ideological progressionism and Darwin's scientific initiative may have catapulted pseudo-Darwinism to prominence, but conservative forces were quick to adapt, and they retained an influence on evolutionary thought in the later decades of the nineteenth century.

Even the most hostile critics of Darwinism conceded that some aspects of evolution are adaptive, but there was widespread reluctance to admit that natural selection operates even on a reduced scale. Moral and theological objections played a vital role in generating enthusiasm for a non-Darwinian interpretation of adaptation. Would a wise and benevolent God find it necessary to kill off so many unfit individuals to ensure the breeding success of the few lucky enough to inherit useful characters? Spencer had exploited Lamarckism to give a more positive image in which the majority of individuals learn to cope with environmental challenges and pass on their acquired characters. This was consistent with the Protestant ethic's emphasis on the benefits of hard work and enterprise, but Spencer's largely materialist approach to psychology did not encourage his followers to stress the element of creativity inherent in the Lamarckian assumption that behavioral initiative lies at the heart of evolution. Spencer saw the acquisition of new habits and new physical adaptations as a more or less automatic process. But if one chose instead to stress the act of "discovering" a new behavior pattern to cope with a changed environment, it would be possible to see the inherent creativity of living things as a divinely implanted force that allowed them to triumph over the material world.

Opponents of materialism could thus use Lamarckism as the basis for a new solution to the problem of design. The struggle and suffering of natural selection were rejected as unnecessary. Instead, the initiative shown by organisms responding to their environment could be seen as a divine spark of creativity implanted into nature to serve as the guiding force of evolution. Instead of portraying the individual as a puppet at the mercy of biological forces, this vitalist interpretation of Lamarckism stressed the freedom of choice open to organisms as they sought to confront the external world. Many religious thinkers welcomed the idea that God has somehow delegated His power of creativity to the life force of the individual, so that species can redesign themselves through the evolutionary process. This interpretation of neo-Lamarckism was promoted by Edward Drinker Cope and other members of the American school from the 1870s onward (Moore 1979, chap. 6). In the long run, however, it achieved greater prominence through the writings of the novelist Samuel Butler (Willey 1960; Bowler 1983, chap. 4).

It was Mivart's *Genesis of Species* that first encouraged Butler to challenge Darwinism, and he later wrote to Mivart arguing that it was better to see the Designer's power internalized within nature than to abandon Him altogether (Jones 1919, 1:407). In a series of books, beginning with *Evolution, Old and New* (1879), Butler lambasted both Darwin's character and the materialism of his theory. In response to Weismann's neo-Darwinism he wrote, "To state this doctrine is to arouse instinctive loathing; it is my fortunate task to maintain that such a nightmare of waste and death is as baseless as it is repulsive" (repr. in Butler 1908, 308). Although at first dismissed by the British scientific community, Butler enjoyed some popularity during the temporary surge of enthusiasm for anti-Darwinian ideas in the 1890s. These ideas soon began to vanish from biology in the early twentieth century, of course, but they were still vigorously promoted by literary figures such as Bernard Shaw (in the preface to

Back to Methuselah) and, more recently, Arthur Koestler (1967.)

Butler's attack on Darwinism reminds us that literature may provide valuable insights into how the Victorians dealt with the religious and moral implications of evolutionism. Minor novelists were quick to seize upon the loss of faith provoked by evolutionism as a theme to illustrate the tensions in society. The survey by Leo Henkin (1963) revealed that such novels soon disappeared from the scene as the controversy died down. Satires of Darwinism also became less common after 1880 as novelists realized that a compromise had been reached between evolution and religion. Perhaps more interesting is the way in which metaphors and themes associated with Darwinism found a place in literature (Hyman 1962; Beer 1983; Leatherdale 1983). George Meredith is perhaps the best example of a novelist who adopted an optimistic progressionism more in line with pseudo-Darwinism. But Darwin's use of metaphors such as the "tangled bank" to reflect the complexity and indeterminacy of organic relationships seems to have struck a responsive chord, especially in George Eliot and Thomas Hardy. Hardy also accepted a harsh image of the individual at the mercy of heredity and of unbending natural laws (Beer 1983; Morton 1984).

The vision of nature proclaimed in the *Origin of Species* was thus capable of stimulating some of the most imaginative Victorian writers. But however interesting these parallels may be to the literary historian, it must be remembered that great novelists do not merely reflect the general culture of their time. Precisely because they are creative artists, they are capable to some extent of transcending the culture in which they live. As Ellegård (1958, 23) pointed out, the routine periodical literature of the Victorian era may be less exciting than the great novels, but it is a better indication of contemporary thought. Measured by Ellegård's yardstick, Darwinian values did not penetrate very far into the culture of the 1860s and 1870s, and they only became commonplace among lesser writers toward the end of the century. American naturalistic writers such as Jack London eventually began to portray the world as a scene of constant struggle (Russett 1976, chap. 7). In Britain, the science-fiction stories of H. G. Wells began to reflect a sense of evolutionary pessimism during the 1890s (Morton 1984). *The War of the Worlds* extended the struggle for existence onto an interplanetary scale, with mankind very nearly losing out to the Martians. *The Time Machine* depicted a future in which the human race has stagnated because technology has eliminated the need to struggle against nature. Wells's rejection of progressionism is usually attributed to his contact with T. H. Huxley, whose

"Evolution and Ethics" had invoked a "cosmic pessimism" based on the view that evolution is not directed toward a moral goal (Philmus and Hughes 1975).

If popular novels are any indication, then, it was not until the 1890s that the English-speaking world began to exhibit a general willingness to doubt the inevitability of progress. Far from creating an image of nature as a meaningless cycle, both pseudo-Darwinism and the various forms of anti-Darwinism had succeeded in retaining significant aspects of the traditional value system. Evolutionists might hail the Protestant work ethic with Spencer, the divine plan of creation with Mivart, or the spontaneity of individual choice with Butler, but all used the idea of purposeful or progressive change to retain a faith that nature itself works according to moral principles. The crisis of faith had simply evaporated, because progressionism allowed liberal and conservatives alike to feel that the most threatening implications of Darwin's theory had been evaded. Even in the 1890s, pessimism still ran only skin deep. There was certainly a *fin de siècle* concern for the threat of cultural degeneration, presented by writers such as Oscar Wilde (see Nordau 1895). Yet recent studies of this movement have suggested that decadence was seen only as a temporary threat. Hardly anyone doubted that progress would resume once suitable conditions were restored (Chamberlin and Gilman 1985). Only in the early decades of the new century did major segments of Western culture begin to suspect that progress was an illusion soon to be swept away by the triumph of irrationality and brute force. Here at last was the real crisis of faith, which would shatter the Victorians' optimism by destroying the compromise through which the earlier, less acute crisis had been resolved.

But what role did Darwinism play in this cultural revolution marking the transition from nineteenth- to twentieth-century thought? The writings of Huxley and Wells certainly indicate that recognition of a genuinely Darwinian view of undirected evolution could undermine faith in progress. The carnage of World War I became the most obvious symbol of the West's moral failure, and "social Darwinian" attitudes were subsequently blamed for creating the tide of hate that led to war. But Darwin's theory was *not* popular among biologists and sociologists in the prewar years, and it is difficult to believe that a largely discredited theory could have played a major role in whipping up enthusiasm for conflict. At best, the memory of Darwinism may have created a loose framework of metaphors by which attitudes derived from other sources could be justified. The growing sense that human nature is governed by irrational impulses was fueled by a new generation of iconoclastic artists and by psychologists such as Freud. To see if Darwinism played a role in the destruction of progressionism, I must now evaluate its influence upon those thinkers who established the new directions that would be taken by twentieth-century thought.

Darwin and Modern Culture

The Victorian crisis of faith, with its recognizable peak in the 1860s, was very much an affair of the English-speaking world. In continental Europe, the process of secularization had its own crises, but it ground on inexorably throughout the century. Yet the secularized world had not necessarily abandoned traditional values; it merely reinterpreted them. Europe had its atheists and materialists, perhaps more prominent than any in Britain and America, but the popularity of philosophies such as Haeckel's monistic developmentalism ensured that here, too, there was wide support for a world view still unified by a sense of purpose in nature. Western culture as a whole seems to have reached a crossroads at the beginning of the twentieth century. Americans and Europeans alike suddenly found themselves doubting the moral foundations they had used to support their domination of both the physical world and the rest of the human species. They could no longer be sure that nature intended them to rule the world, nor could they trust the rational thought-processes that had once seemed to guarantee their superiority.

The chaos of political, moral, and artistic philosophies that seems to characterize modern thought stems from this loss of faith in the power of reason to uncover the orderly foundations of natural development. Nor is the chaos confined to the philosophical level. Extremist ideologies have been implemented again and again in the twentieth century, often with disastrous results. Liberal democracies have survived the transition from military to economic imperialism, but they have found their authority challenged by movements of both the Left and the Right. Curiously, it is these extremes that have best preserved the confidence of nineteenth-century developmentalism. Marxism has retained a sense of the inevitability of progress toward the socialist paradise and has sustained its supporters through a succession of hollow victories. It was able to do this by internationalizing the process of development, so that all races could participate, at least in theory. On the Right, the Fascists and Nazis glorified the struggle by which the superior race would dominate the weak, using the image of the great leader who expresses the aspirations of his people. It is the center that has suffered the loss of confidence that seems more characteristic of modern thought. As W. B. Yeats wrote in "The Second Coming," just after World War I,

Things fall apart; the centre cannot hold;
Mere anarchy is loosed upon the world,
The blood-dimmed tide is loosed, and everywhere
The ceremony of innocence is drowned;
The best lack all conviction, while the worst
Are full of passionate intensity.

My task is to assess the extent to which Darwinism contributed to this fragmentation of cultural and political life. Was the loss of faith in progress due to a belated recognition of the threat posed by Darwinism to the developmentalist compromise? Or were other forces, with separate origins in nineteenth-century thought, of greater influence? The examples of T. H. Huxley and H. G. Wells cited above suggest that Darwinism may have played some role in arousing the doubts of the *fin de siècle* period. The general idea of evolution certainly remained a fundamental component of many trends within the new cultural ferment. But I shall argue that the specifically Darwinian component of evolutionism — as understood by those who have raised the theory to its present status in biology — was of relatively minor importance. The emergence of Darwinism as a major force in modern biology should not be taken as a sign of the theory's more general cultural influence, partly because biology itself has lost much of its power to dominate our view of human nature and society.

It is a commonplace enough observation that the realities of political life under a regime of the extreme Left or the extreme Right are often very similar. In both cases the individual loses the freedoms so highly prized by liberal democracies and is submerged into a unified state that demands absolute obedience so that some future goal can be achieved. Jacques Barzun's *Darwin, Marx, Wagner* ([1941] 1958) tried to associate Darwin with this dehumanizing trend, presenting him as a leading figure in the move to subordinate the individual to universal forces. This point may have some validity for pseudo-Darwinism, but fails to take note of the fact that modern biologists have welcomed Darwin's insights precisely because they do *not* imply that the course of future development is predetermined. In a world where a species' evolution is subject to the hazards of environmental change, no one could claim by analogy that social progress is predetermined. If Darwin's name is to be linked to the Right's fascination with national or racial domination, it must be through Haeckel's pseudo-Darwinian developmentalist philosophy of nature. Daniel Gasman (1971) argued that the social Darwinism of Haeckel's Monist League influenced a number of Nazis at an early stage in their careers. Gasman's view of Haeckel's politics has been challenged (e.g., Kelly 1981), but Gasman has defended his position by pointing to some very strange developments in Haeckel's later life (see below).

Later perceptions of Darwin's work frequently do not reflect the reality. Historians concerned with the real Darwin, the real Haeckel, or the real Nietzsche may complain about the distorted character of their hero's message when proclaimed by the madmen of the Third Reich. But the extreme Right has borrowed images eclectically from all sides, and its exaltation in the "struggle for existence" may be no more authentic than its worship of the "superman." The notion of struggle as a vehicle for progress had many roots in nineteenth-century thought, and the fact that two such dissimilar thinkers as Darwin and Nietzsche can be regarded as its originators confirms only that neither had any control over the images created by his ideas in the popular mind.

At the opposite end of the political spectrum, Marx welcomed the *Origin* as a valuable foundation for materialism (Heyer 1982). As he wrote to Engels in 1860, "Although it is developed in the crude English style, this is the book which contains the basis in natural history for our view" (Marx and Engels 1953, 170). Evolutionism was certainly important to Marx, but, as I have noted (chap. 2 above), his complaint about the English style of the *Origin* included a recognition that natural selection was a reflection of the free-enterprise system. The once-popular assumption that Marx offered to dedicate a volume of *Das Kapital* to Darwin is now known to be based on a misinterpretation of the relevant correspondence (Colp 1974; Feuer 1975, 1976; Fay 1978). Popular writers still occasionally suggest that the Marxist concept of class struggle is derived from the Darwinian struggle for existence (e.g., Burke 1985, 273, and more strongly in the TV program), but this is surely a misconception. Marx's view of evolution was openly developmental — his favorite illustration was the growth of a seed — and his interest in struggle emerged from the dialectical view of development he inherited from Hegel. The Lysenko affair in the Soviet Union confirmed that Marxism can easily be used as the grounds for rejecting both Mendelian genetics and Darwinian selectionism (Medvedev 1969; Joravsky 1970).

If Marxism preserved the belief that human activity must in the end contribute to a universal progressive trend, the Nazis' sense of racial superiority was characterized by irrational elements that constitute some of the most disturbing elements of modern thought. We live amid a host of rival interpretations of human nature, some of which deliberately turn their backs on the nineteenth century's hope that reason must prevail because it is in harmony with nature. We are told that the human mind is open to domination by whatever cultural or social system is imposed upon it, or that its capacities and thought-processes are rigidly determined by inheritance or the subconscious relics of animal instincts. In various ramifications, the nature-nurture debate still lies at the heart of our modern uncertainty about our own essence. But what role did Darwinism play in shaping this modern dilemma?

Social scientists and psychologists offer a bewildering variety of opinions on the essential nature of the human mind and its suscep- tibility to biological and cultural influences. The social sciences of the early twentieth century proclaimed their independence of bio- logical evolutionism and began to treat culture as the product of an autonomous human capacity (Cravens 1978; Greenwood 1984). The modern anthropologists' acceptance of cultural pluralism mimics the Darwinians' rejection of the developmental model but owes little to a Darwinian analysis of how change takes place. Cultures evolve through a process that is more closely analogous to Lamarckism, since it involves the "inheritance" of innovations through learning. Social scientists are convinced that the human mind is predisposed to accept culture, but not a particular type of culture, and they often show little interest in the paleoanthropologists' efforts to explain how the cul- tural capacity may have evolved in our early ancestors.

In psychology, the closest equivalent to the social scientists' break with biological determinism was the emergence of J. B. Watson's behaviorism (Boring [1929] 1950, chap. 25; Cravens 1978, chap. 6; Boakes 1984). Launched in 1913, Watson's movement challenged the prevailing view that behavior is governed largely by instinct. He pre- sented an image of the animal and the human mind as infinitely malleable under the pressure of external stimuli. Enhanced by rein- forcement (gratification of immediate biological needs), the learning process could instill any kind of conditioned reflex into the individ- ual's behavior pattern. Behaviorism took evolution for granted in asserting the equivalence of human and animal activities, but it elim- inated the concept of "mind" from scientific discourse. The most obvious biological influence, the desire to study behavior along ex- perimental lines identical to those used in the investigation of other organic processes, came from physiology rather than evolutionism. The broader implications of behaviorism were not lost on Watson or his followers. If their picture of human nature was valid, people could be conditioned to live happily in a society designed by the state. Such a view led directly to the *Brave New World* explored not only in Aldous Huxley's novel but also in B. F. Skinner's *Walden Two* (1948).

Behaviorism seemed to destroy the individual's autonomy by reducing him or her to a puppet in the hands of whoever controls the external stimuli. Watson and his followers were reacting against a theory that equally constrained the human mind, but did so through rigidly inherited instincts. William MacDougall popularized the instinct theory in his *Introduction to Social Psychology* of 1908, citing evolution theory to explain the origin of human instincts. Although at first prepared to consider Darwinian models, MacDougall eventually turned to Lamarckism (1927), and it is probable that Lamarckism contributed far more to the popularity of the original theory. The concept of instinct as inherited habit died very slowly, whatever the evidence from biology.

The heyday of instinct theory in psychology coincided with the early years of the eugenics movement, which assumed that mental ability is determined by biological inheritance (chap. 7 above). Psychologists were called in to develop the intelligence tests needed to identify the feebleminded individuals who should be institutionalized or sterilized (Gould 1981; Evans and Waites 1981). The eugenics movement reveals the loss of confidence which began to affect Western nations in the early twentieth century. Far from accepting nature as a guarantee of success for the virtuous and creative aspects of human nature, eugenics saw unrestrained breeding as a threat that would swamp society with ever greater numbers of the unfit. This seems to reflect the basic Darwinian point that reproductive success alone determines the biological future of the race. As I have noted, however, eugenics was supported by many of the early Mendelians, who did not believe that natural selection was the driving force of evolution. It may be significant that both Mendelism and behaviorism were inspired by a desire to extend the experimental method into the study of life. In biology, at least, the theory of heredity which was established on the basis of this new method far outweighed any lingering influence of Darwinism as a means of undermining the old developmentalist viewpoint. On a broader front, however, the extreme environmentalist and hereditarian positions merely added to the confusion. However dramatic its effect on biology, no historian would think of the Mendelian Revolution as a turning point in the emergence of modern thought.

Eugenics assumed that success in reproduction was not related to intellect but still held out hope that rational control of human breeding could avert degeneration. Far more damaging to the West's self-confidence was the recognition that even the noblest intellect could be influenced by the subconscious forces of sexuality. Sigmund Freud himself regarded his discovery of the subconscious mind as the last of three great blows to humanity's self-esteem, following those administered by Copernicus and Darwin (Freud 1959–74, 16:284). In a very important sense, Freud's own revolution depended upon Darwin's; to believe that human behavior is conditioned by sexual factors, one must accept that human beings are little more than highly developed animals. Frank Sulloway's detailed study of the biological origins of Freud's thought (1979b) revealed the extent to which the founder of psychoanalysis built his view of the mind on evolutionary foundations, although his followers have repudiated the connection.

But was it a *Darwinian* foundation? Although Sulloway included a chapter entitled "The Darwinian Revolution's Legacy to Psychology and Psychoanalysis," it is clear that for the most part Freud's biology was at best pseudo-Darwinian and often openly Lamarckian. His belief that the various levels of the mind conflict with one another may be yet another dubious extension of the struggle metaphor, but Freud was most deeply impressed by the recapitulation theory and the inheritance of acquired characters. Sulloway showed how earlier psychologists such as G. J. Romanes, J. M. Baldwin, and G. Stanley Hall adopted a recapitulationist approach to the growth of the individual mind (see also Gould 1977b, chap. 5). They were led to suspect the early origin of sexual feelings, since sex itself appeared at a very early stage in the history of life on earth. Freud's whole conception of the various levels of the subconscious mind reflects a hierarchical theory of development which posits the retention of animal-like instincts despite their being overlaid by higher functions. He also adopted a Lamarckian explanation of how experiences become imprinted on the racial memory, dismissing the increasing skepticism of early-twentieth-century biologists (Sulloway 1979b, 373, 386–88, 439–40). It was the developmental, not the Darwinian, view of evolution that played the most vital role in the origin of this most disturbing facet of modern thought. (It may be added that Lamarckism was also a strong influence upon Jean Piaget, whose genetical epistemology has played a less controversial role in shaping modern ideas about the human mind; see Piaget 1979; Rotman 1977.)

In his *Discovery of the Unconscious* (1970) Henri Ellenberger linked Freud's ideas not to Darwin but to Ernst Haeckel's recapitulationism, perhaps the most blatant late-nineteenth-century expression of the analogy between growth and evolution. More recently Daniel Gasman (abstract 1985) stressed the broader influence of the vitalism and mysticism that lay behind the facade of Haeckel's materialism. The monist philosophy expounded in his immensely popular *Riddle of the Universe* (trans. 1900) accepted, in effect, that matter is alive and proclaimed the pantheistic belief that the whole of nature is a unified spiritual entity. These points link Haeckel back to the idealist and transcendentalist origins of the developmental view of nature and confirm the teleological character of his progressionism. Freud's discovery of the unconscious mind can thus be seen as the last, self-destructive expression of the developmentalist thesis that the human mind is the end product of a unified natural process.

Gasman also showed how Haeckel's philosophy was seized upon by a wide range of thinkers not usually associated with the sciences. Haeckel himself had contacts with the Theosophist Rudolph Steiner and seems not to have discouraged the view that his work could be used as the foundation for a new, non-Christian mysticism. Among the writers and artists who began to express dissatisfaction with the triumphs of Western civilization at this time, there were many who had been influenced by the mystical aspects of monism, including the painters Paul Klee and Vasily Kandinsky and symbolist writers such as August Strindberg and Franz Kafka. The primitivism that influenced so many early modern artists may also have been an expression of the monistic sense of unity between mind and nature. Like Freud, these artists saw the human mind as having deep roots in nature that may be more powerful than the superficial trappings of rationalism. If Gasman's interpretation is accepted, it becomes necessary to reconsider the superficially obvious view that the loss of faith in rationalism was prompted by a materialist (and hence Darwinian) sense that humans are no more than animals. Some of the most revolutionary innovations of twentieth-century thought may now have to be seen as the final expressions of a cultural trend arising from the most explicitly antimaterialist currents responsible for the creation of the developmental world view.

From the start, some of the artists who appreciated the primitive foundations of the human mind shared Freud's concern that irrational impulses may be highly disruptive. The horrors of World War I brought home to everyone the self-destructive forces lying at the heart of Western civilization and made it possible to believe that those forces are inherent in the human personality. Soon many thinkers, of whom the existentialists are perhaps the best-known examples, began to stress the individual's alienation from society and from the world itself. Such attitudes rested upon a new vision, not just of the human personality but of the world in which it is forced to live. Here was a second revolution undermining the traditional faith in nature as a guarantee for the triumphs of rationality and of Western culture. The possibility that nature itself might be amoral and purposeless had been expressed by Huxley, so a Darwinian influence on this second revolution cannot be ruled out completely. But even here, there were many other factors at work. Henry Adams, whose superficial brush with the Darwinism of the 1860s has already been noted, was acutely aware that the turn of the century was marked by an intellectual revolution. The key breakthrough, he felt, was a sudden recognition of the possibility that nature is a system without order or purpose: "Chaos was the law of nature; Order the dream of man" (Adams 1918, 451). Nineteenth-century evolutionists such as Haeckel represented the apogee of a long-established tradition in which mankind had attempted to impose a sense of order upon nature. Adams now saw this tradition being broken—not through an admission of the primitive nature of the human mind but by a demonstration of nature's indifference to human values.

Adams identified two major influences that had forced him to confront this situation. The first was Karl Pearson's *Grammar of Science* ([1892] 1900), which advocated a positivist stance, dismissing all the unifying factors of nineteenth-century science—order, reason, beauty, and benevolence— as products of the human mind with no foundations in nature itself. "Suddenly, in 1900, science raised its head and denied" (Adams 1918, 452). Pearson was, of course, one of the few prominent Darwinists of the time, and it is possible that the Darwinian view of nature had helped to stimulate his rejection of the earlier tradition. But Adams made no mention of this; for him the most important confirmations of the new view of the world had come from the physical sciences. The discoveries of Roentgen and the Curies had proclaimed the chaotic state of the natural world. The "metaphysical bomb called radium" had led Arthur Balfour and others to declare that "the human race without exception had lived and died in a world of illusion until the last years of the century" (Adams 1918, 452, 457). Adams understood the details of the new atomic physics no better than he had once understood Darwinism, but he sensed that these discoveries had somehow undermined the traditional faith in the stability and orderliness of natural processes.

The cultural shock administered by atomic physics was reinforced by Einstein's discovery of the principle of relativity. Again, few understood it, but everyone felt that the foundations of the old order of nature had been threatened. Einstein was elevated to the status of a cult figure— the prophet who proclaimed that all values are merely relative and that nothing in this world is really certain (Friedman and Donley 1985). Einstein himself did not believe that God plays dice with the universe, and there were some efforts to develop a more comforting view of nature's interrelatedness with the human mind (e.g., Alfred North Whitehead 1929). But many shared Adams's fear that science had now revealed a world of frightening instability. These fears may seem trivial now (given the real threats posed by the technological spinoffs from the new physics), but they clearly served to convince many early-twentieth-century thinkers that the foundations of the traditional world view had been shattered.

The developmental view of evolution lost credibility in biology through the emergence of Mendelian genetics, although it lingered on for a while in areas such as paleontology. On a broader scale, however, this revolution was but one among many that were responsible for generating a sense that nature is a purposeless system that does not move in a direction likely to coincide with human moral values. Whatever the origins of this new climate of opinion, Darwinism now remained the biologists' only hope for showing some rough-and-ready order in nature's haphazard activities. Elimination of developmentalism cleared the way for naturalists to reassess the geographical factors that had so impressed Darwin and for geneticists to realize that selection is the only mechanism capable of directing a sequence of random mutations. The open-ended and indeterminist character of Darwin's theory was ideally suited to the thought-processes of the new century. Although largely incompatible with the values of Darwin's own time, natural selection could now take its place as a leading scientific innovation. Yet it is hard to see the emergence of modern Darwinism as anything more than a byproduct of this later cultural revolution. The theory's role in the development of twentieth-century thought has been largely peripheral, unless the advent of sociobiology forces a major change in the attitudes of psychologists and social scientists. It thus seems unreasonable for historians to claim that the turning point in the emergence of modern culture should be called a "Darwinian Revolution."

Chapter Nine

TOWARD A NEW HISTORIOGRAPHY
OF EVOLUTIONISM

✤ In 1986 two of the most eminent historians of evolutionism crossed swords over the origin and significance of Darwinism (Greene 1986; Mayr 1986). John C. Greene, whose work on the cultural impact of evolutionism has inspired many younger historians (myself included), presented Darwin as an aberrant product of the materialist way of thinking which first began to gain ground in the eighteenth century. Greene saw Darwin's ambiguous support for progressionism as an indication of materialism's inherent weakness; this philosophy can only triumph by temporarily pretending that blind nature does, after all, have a purpose. Ernst Mayr, whose position as a leading modern Darwinist serves as the foundation for his study of evolutionism's scientific origins, responded to Greene by insisting that Darwin pioneered a new and fruitful way of showing how the routine operations of natural law can achieve the results once attributed to divine forethought.

Clearly Greene and Mayr disagree fundamentally over the value of materialism as a philosophy, but their perspectives on Darwinism have also been shaped by their very different historical priorities. Mayr is concerned with the explication of Darwinism's success as a scientific theory, while Greene worries about its role in undermining the religious foundation that once underpinned Western culture. As a result they find it difficult to communicate with one another; they use vocabularies in which the term *Darwinism* has different meanings. Yet on one point they *do* agree. Both seem content to accept that the rise of evolutionism should be understood primarily as a contribution to materialist thought. Neither makes any serious reference to the role played by developmentalist attitudes in shaping the nineteenth-century origins of modern evolutionism. It is precisely this one-sided viewpoint that I have set out to challenge in the chapters above, and I now want to argue that only by taking non-Darwinian factors into account will these two scholars enable their divergent perspectives to interact fruitfully with one another. Both have exaggerated the role of Darwinism, but because they have done so for very different reasons they are unable to agree on the significance of the theory that looms so large in their deliberations.

The contention of this book is that the orthodox image of the Darwinian Revolution, still visible in the debate between Greene and Mayr, has distorted the perception of how evolutionism originated and hence the view of its effect on modern thought. The acceptance of natural selection by modern biologists has led to an overemphasis on Darwin's role in the history of nineteenth-century evolutionism and a failure to appreciate the extent to which the topic has been interpreted in a non-Darwinian way. Darwin's more radical ideas seem important to modern biologists, but they were uncharacteristic of his own time; they served largely as a catalyst, helping the supporters of developmentalism to create a fully evolutionary world view that preserved the belief in an orderly and purposeful nature. Darwin certainly supplied images and metaphors that appealed to the Victorian mind, but the concept of "progress through struggle" was more often understood in a Lamarckian context that bears little relationship to the parts of the theory remembered by modern biologists. Even Darwin only occasionally realized that he had created a system capable of challenging the progressionist spirit of the age. The thoroughgoing materialism of the selection theory was at first proclaimed only by conservatives anxious to discredit evolutionism in general, although in later decades the Lamarckians took on a similar role. The *Origin* debate triggered a growing wave of support for developmental evolutionism that passed from a pseudo-Darwinian into an anti-Darwinian phase as the century drew to a close.

With this model, the growth of nineteenth-century evolutionism can be seen as a unified movement toward a progressionist world view rather than as two distinct episodes separated by the publication of the *Origin*. Darwin's thinking drew upon contemporary ideas and values but introduced a significant twist that would not be properly appreciated until those values were swept away in the twentieth century. The appearance of the *Origin* acted as a stimulus to the growth of developmentalism, but it did not precipitate a breakthrough into full-scale materialism. Historians certainly need to tackle the question of why this stimulus was required to encourage the evolutionary potential of developmentalism to become manifest. But the growing recognition of the role played by non-Darwinian models in the post- as well as the pre-*Origin* period undermines the plausibility of the traditional picture in which the publication of Darwin's theory represents the only major watershed in the rise of evolutionism. Instead of portraying Chambers's *Vestiges* as primitive "pre-Darwinian" evolutionism and the neo-Lamarckians as misguided "anti-Darwinians," historians should see them as merely the earlier and later phases of a continuous process of conceptual development, a process accelerated but not seriously distorted by the publication of Darwin's work. The revolution at the end of the century caused (in biology, at least) by the advent of Mendelian genetics will then take on a greater significance because it — not the temporary disturbance of the *Origin* debate — marks the decisive break with developmentalism, and hence with teleological thinking. I have noted above that, outside biology, developmentalism itself helped to sow the seeds that would grow into some of the more nihilistic aspects of twentieth-century thought.

Historians studying the cultural impact of evolutionism have already paved the way for a growing recognition of the role played by non-Darwinian concepts. If "Darwinism" promoted a revolution within, not against, natural theology (see chap. 8 above), then it follows that the implications of the selection mechanism spelled out so clearly by modern biologists could not have been appreciated at the time. Pseudo-Darwinism required a massive compromise with progressionism, while anti-Darwinian theories often encouraged an openly teleological view of nature and of human origins. Far from treating Darwin as an anomalous materialist, it might be better to see him as an anomalous developmentalist whose theory contained implications that would only be fully realized after the emergence of modern hereditarianism. Cultural historians must recognize that "Darwinism" is now loaded with connotations established during the triumph of the synthetic theory within modern biology. Using "social Darwinism" and other such expressions to denote late-nineteenth-century attitudes will inevitably mislead anyone whose image of Darwinism is shaped by the modern, post-Mendelian understanding of how selection works. Social *evolutionism* contained some Darwinian, but also many non-Darwinian, elements, and only by bringing this fact squarely into assessments of late-nineteenth-century culture will it be possible to ensure that misunderstandings do not arise.

This does not mean that an attempt should be made to hide the cultural and ideological significance of the evolutionary philosophies promoted by Spencer, Haeckel, and others. They certainly encouraged a callous indifference to the fate of those individuals or races that could not keep up with the march of progress. A new way of justifying social policies had entered Western consciousness, and there is a continuous process linking nineteenth-century "Darwinism" to modern ideas of genetical determinism and sociobiology. But emphasizing the Darwinian component of social Darwinism only interferes with an appreciation of the complex and changing role played by biology in the course of the last century or more. Early forms of biological determinism (including race theory) owed a great deal to non-Darwinian concepts, and unless historians trace the way in which those concepts were replaced by more rigidly hereditarian views in the twentieth century, the foundations upon which modern attitudes have been built will continue to be misunderstood. Using Darwin as the figurehead of nineteenth-century evolutionism is understandable in view of the dramatic effect of his theory in precipitating the general conversion to a world view based on continuous development. But it is dangerous in that it encourages the assumption that his contemporaries were attracted by the implications that modern biologists have extracted from his theory. Historians seeking to understand the rise of materialism will do well to recognize that Darwin's theory cannot be made the scapegoat for all subsequent developments in that area. Darwinism itself was absorbed into nineteenth-century progressionism, and non-Darwinian values contributed far more than is normally realized both to the first wave of determinist ideology and to the foundations of modern thought.

The temptation for cultural historians to use Darwin's theory as a symbol of changing attitudes in the nineteenth century has been encouraged by the historians of biology, whose attention has been focused on the discovery of natural selection because this theory is seen as the key to the emergence of modern scientific evolutionism. Recognition of the role played by non-Darwinian theories will affect the history of biology, too, but in a less direct manner. To some extent, biologists such as Mayr have an excuse for ignoring the extensive network of non-Darwinian thinking in the late nineteenth century. As modern Darwinians, they are bound to take a greater interest in the thin red line running from Darwin to the genetical theory of natural selection. One cannot altogether blame the Whigs for writing Whig history, especially when, as in the case of Mayr, they freely admit the selective nature of their interest in the past. The history of science must allow room for those whose primary concern is the use of hindsight to uncover the origins of modern theories. Only when implemented in an unsympathetic manner does this concern for the "main line" of scientific development lead to a positive distrust of historians who wish to revive interest in long-forgotten blind alleys. And yet some scientists who criticize historians for studying now-rejected theories are apparently under the impression that the historians are promoting a relativistic view of knowledge in which one theory is as good as any other. It must be confessed that some radical historians do imply that successful theories achieve their dominant status only because they satisfy ideological or professional interests, not because they describe nature more accurately. But the majority of historians of science, myself included, have no intention of promoting a relativistic philosophy of knowledge, although we do insist that science be seen as a human activity. I hope to persuade even modern biologists that they have something to gain from adopting a more flexible view of how evolutionism developed.

Biologists who believe that the synthesis of Darwinism and genetics will continue to serve as the basis for all future evolutionism may tend to dismiss non-Darwinian theories as of little interest, even if it can be shown that those theories exerted considerable influence in earlier decades. But it may turn out that better knowledge of the "eclipse of Darwinism" will help to solve problems that would otherwise puzzle those who prefer to concentrate on the main line of development revealed by hindsight. A classic example is the perennial question of why Mendel's work was ignored until its "rediscovery" in 1900. Considering the traditional view that Mendel's laws should have been seen as an obvious solution to the problems of Darwinism, the long neglect of his work is something of an enigma. In chapter 5 above I discussed how a broader understanding of late-nineteenth-century evolutionism helps to explain why the link between Darwinism and Mendelism could not be made until later. Mendel's experiments were ignored because the developmental world view prevented anyone from realizing that a study of heredity could be of major importance even though it did not address the question of growth. In other words, we are in a better position to understand the complexity of the main line linking Darwin to modern Darwinism if we are aware of the role played by non-Darwinian (and non-Mendelian) factors.

At the level of public relations, the supporters of modern Darwinism also have much to gain from encouraging better awareness of the role played by alternative theories. Whatever the status of evolutionism within biology, it has never been completely accepted by the population at large, and opposition has grown apace in recent decades. Those who are hostile to Darwinism, including, most obviously, the creationists, frequently claim that the selection theory has always been imposed as a rigid dogma, enforced by the materialist preconceptions of the scientific community. The traditional image of the Darwinian Revolution simply reinforces this claim, which can only be disproved by showing that alternatives to Darwinism have not always been ignored, and that earlier biologists made serious efforts to reconcile evolutionism with teleology. Far from being a long-established dogma, Darwinism only triumphed in the 1940s after many decades in which its rivals had ample opportunity to influence the course of scientific biology.

Stephen Gould (1987) illustrated how a greater awareness of the role played by non-Darwinian evolutionism can benefit scientists trying to defend the modern theory against its opponents. He noted that there is a historical dimension to the claim by Fred Hoyle and C. Wickramsinghe (1986) that the impressions of feathers were engraved on the famous *Archaeopteryx* fossils to create a fraudulent link between reptiles and birds. To explain the reception of the "fraud," Hoyle appealed to the traditional image of Richard Owen as a bigoted anti-evolutionist. Gould has shown that this image is a caricature; Owen did believe in evolution, but this fact has been obscured by the assumption that anyone who opposed Darwinism must also have rejected evolution itself. Better knowledge of Owen's non-Darwinian evolutionism would undermine the misrepresentation of his position which lends a spurious air of plausibility to Hoyle's claims.

There is one final point that should give modern biologists food for thought. In the last decade or so it has become obvious that there is no longer a universal consensus in favor of the synthetic theory even within the ranks of working biologists. The most loyal supporters of Darwinism must concede that there are other biologists who are not quite so sure that the theory will maintain its hold in the future. In a surprising number of cases, the alternatives that come under discussion seem to recall at least some aspects of the non-Darwinian ideas that circulated in the pre-synthesis era (for a brief survey of the modern debates see Bowler 1984a, chap. 12). The popular image of evolutionism has never adapted completely to the nondevelopmental approach of synthetic Darwinism, but it now seems possible that biology itself may be forced to acknowledge growth-factors as an element that must be incorporated into future theories. Significantly, the biologists who question the adequacy of Darwinism have different perspectives on the history of evolutionism. Some (e.g., Løvtrup 1987) adopt an extravagant hostility to Darwin's contributions which can hardly be condoned. But at a more reasonable level, Niles Eldredge (1986) uses his support for punctuated equilibrium theory to question the orthodox histories of Darwin's achievement on the grounds that they ignore his failure to confront the evident discontinuity of the fossil record. The questions we ask about the past reflect our own interests in the present. Biologists who advocate a pluralistic approach to modern evolutionism ought to welcome historical studies that loosen the hold gained by the "Darwinian Revolution" upon our imaginations.

What, then, will be the role of the Darwin industry in the new historiography of evolutionism? It would be wrong to assume that a greater appreciation of the role played by non-Darwinian factors will produce an exactly equivalent diminution in the amount of effort expended on Darwin himself, although, clearly, some reassessment of priorities will be necessary. The arguments presented in this book have drawn upon the efforts of those scholars who have devoted so much of their time to a minute analysis of Darwin's papers and notebooks. Far from merely filling in the details of the traditional image, the Darwin industry has helped to generate a new interpretation in which natural selection (in its original form) is seen to be deeply imbued with a generationist view of heredity and variation. Such revolutionary insights were probably not anticipated by the scholars who first began to edit and publish the Darwin papers, but they fully justify the work that has gone into the project. The ongoing plan to edit and publish all of Darwin's correspondence is similarly throwing light on the complex process by which he built up a communications network that would allow his more radical insights to be injected into the debate without provoking an instant rejection of the whole evolutionary program. Those of us whose interests lie in the reception rather than the origins of Darwinism will reap a rich harvest when the post-1859 correspondence appears. Only by studying in detail the relationship between Darwin and his contemporaries can we hope to flesh out (or disprove) interpretations of his role such as those I have suggested above.

The Darwin archives will retain their importance because Darwin remains a central figure in the debate that converted the scientific world to evolutionism. The interpretations suggested here may alter the perception of his role, but they also raise new problems about his interaction with contemporary naturalists and offer new contexts in which his correspondence should be analyzed. At the same time, however, it will have to be recognized that the Darwin papers offer an unevenly balanced view of the debate. They are ideal for answering questions about how Darwin's theory precipitated the transition to evolutionism, but they do not always reflect the growth of non-Darwinian ideas among evolutionists who would have found it increasingly difficult to interact with Darwin and his circle. Darwin is a central figure still, but he should no longer be the pre-eminent focus of attention. Most important of all, we must not let the increasing availability of his papers blind us to the fact that there were developments in nineteenth-century evolutionism in which his participation was not the critical factor.

To those who fear that my call for a new historiography of evolutionism threatens the Darwin industry, I offer a model of what might happen based on recent developments in a related field. Over the last decade or two, the history of geology has been transformed in a way that closely parallels the re-evaluation of Darwinism sketched in above, and yet the subject flourishes as vigorously as ever. At one time it was customary to treat the "uniformitarian-catastrophist" debate as the crucial breakthrough in nineteenth-century geology. Charles Lyell was the heroic figure whose uniformitarian method was supposed to have freed the science from the domination of outdated religious values. Catastrophism was a bad theory that had held back the development of geology with its efforts to defend the reality of Noah's flood and other miraculous interventions (Gillispie 1959). Modern historians of geology have dramatically altered their attitude toward Lyell's position. They now realize that his uniformitarianism contained elements unacceptable to the modern geologist, while the catastrophists made their own important contributions to the current view of the earth's development. This reinterpretation is particularly relevant here because it confirms that the interest in geological catastrophes was partly a byproduct of the deeper concern with establishing a developmental view of the earth's past (Rudwick 1971; Bowler 1976a).

More recently still it has been shown that the uniformitarian-catastrophist debate was of little significance in continental Europe (Greene 1982), while even in Britain the expansion of stratigraphy proceeded without much reference to the debate on geological dynamics (Rudwick 1985; Secord 1986). The great triumph of historical geology in the nineteenth century—the establishment of the sequence of geological periods and their fossil inhabitants—was made possible by the fragmentation of earlier cosmogonic interests in which causal hypotheses had shaped the expected sequence of past events (Laudan 1987). Modern plate tectonics and the theory of continental drift emerged not from this tradition of historical geology but from a revival of interest in the partly neglected topic of geological causation made possible by new techniques for studying the earth's crust (Wood 1985). The parallel with biological evolutionism is not exact; there was never so complete a separation between the causal and historical modes of explanation in evolutionism. Yet it may be significant that the transition from developmentalism to modern Darwinism was made possible by a similar revival of interest in the causes of evolution following the new insights into heredity and variation offered by Mendelian genetics. In geology, as in biology, a new interpretation of the nineteenth century throws greater emphasis onto a second wave of theoretical innovation which took place in the first half of the twentieth century.

The original emphasis on Lyell's achievement can now be seen as the result of an unnatural focusing of attention on a particular set of issues within the British scientific debates. In part this exaggeration was prompted by the modern geologists' decision to adopt Lyell as their hero. But one cannot help feeling that it was also a consequence of the fact that the uniformitarian-catastrophist debate made a convenient prelude to the conventional image of the Darwinian Revolution. Lyell was a kind of John the Baptist, sent to prepare the way for Darwin's final triumph over natural theology and the scriptural view of the past. The development of geology is now seen in more sophisticated terms, and the field retains its interest because there are new kinds of questions to ask. Lyell is an important figure still, but he no longer dominates the scene in so overwhelming a fashion. It remains for historians of evolutionism to make a similar reassessment of their own hero-myth and of the complex issues that have been concealed by the traditional image of the Darwinian Revolution.

REFERENCES

Adams, Henry, 1918. *The Education of Henry Adams: An Autobiography.* Boston: Houghton Mifflin.

Agassiz, Louis, 1842. "On the Succession and Development of Organized Beings at the Surface of the Terrestrial Globe." *Edinburgh New Phil. J.* 33:388–99.

———, [1857] 1962. *Essay on Classification.* Reprint. Cambridge, Mass.: Harvard University Press.

Allan, Mea, 1967. *The Hookers of Kew, 1785–1911.* London: Michael Joseph.

Allen, Garland E., 1968. "Thomas Hunt Morgan and the Problem of Natural Selection." *J. Hist. Biol.* 1:113–39.

———, 1969. "Hugo De Vries and the Reception of the Mutation Theory." *J. Hist. Biol.* 2:55–87.

———, 1975a. *Life Science in the Twentieth Century.* New York: John Wiley.

———, 1975b. "Genetics, Eugenics, and Class Struggle." *Genetics* 79:29–45.

———, 1978. *Thomas Hunt Morgan: The Man and his Science.* Princeton: Princeton University Press.

———, 1986a. "T. H. Morgan and the Split between Embryology and Genetics, 1910–35." In *A History of Embryology,* ed. T. J. Horder, J. A. Witkowski, and C. C. Wylie, 113–46. Cambridge: Cambridge University Press.

———, 1986b. "The Eugenics Records Office at Cold Spring Harbor." *Osiris,* 2d ser. 2:225–64.

Ardrey, Robert, 1966. *The Territorial Imperative.* New York: Atheneum.

———, 1976. *The Hunting Hypothesis.* London: Collins.

Argyll, 8th Duke of (G. D. Campbell), 1867. *The Reign of Law.* London: Alexander Strahan.

Bagehot, Walter [1872] 1971. *Physics and Politics; or, Thoughts on the*

Application of the Principles of "Natural Selection" and "Inheritance" to Political Society. Reprint. Farnborough, Eng.: Gregg International.

Bajema, Carl Jay, 1977. *Eugenics: Then and Now*. Stroudsburg, Pa.: Dowden, Hutchinson and Ross.

Baldwin, James Mark, 1902. *Development and Evolution*. New York: Macmillan.

Balfour, Francis Maitland, 1885. *The Works of Francis Maitland Balfour*. 4 vols. London: Macmillan.

Bannister, Robert C., 1979. *Social Darwinism: Science and Myth in Anglo-American Social Thought*. Philadelphia: Temple University Press.

Barnes, Barry, and Steven Shapin, 1979. *Natural Order: Historical Studies of Scientific Culture*. Beverly Hills, Calif.: Sage Publications.

Barrett, Paul H., Donald J. Weinshank, and Timothy T. Gottleber, 1981. *A Concordance to Darwin's Origin of Species, First Edition*. Ithaca, N.Y.: Cornell University Press.

Barthélemy-Madaule, M., 1982. *Lamarck, the Mythical Precursor*. Cambridge, Mass.: MIT Press.

Bartholomew, Michael, 1973. "Lyell and Evolution: An Account of Lyell's Response to the Prospect of an Evolutionary Ancestry for Man." *Brit J. Hist. Sci*. 6:261–303.

———, 1975. "Huxley's Defence of Darwinism." *Annals of Sci*. 32:535–35.

———, 1976. "The Non-Progress of Non-Progressionism: Two Responses to Lyell's Doctrine." *Brit. J. Hist. Sci*. 9:166–74.

Barton, Ruth, 1983. "Evolution: The Whitworth Gun in Huxley's War for the Liberation of Science from Theology." In *The Wider Domain of Evolutionary Thought*, ed. David Oldroyd and Ian Langham, 261–87. Dordrecht: D. Reidel.

Barzun, Jacques [1941] 1958. *Darwin, Marx, Wagner: Critique of a Heritage*. 2d ed. Garden City, N.Y.: Doubleday.

Bates, Henry Walter, 1962. "Contributions to an Insect Fauna of the Amazon Valley: Lepidoptera: Heliconidae." *Trans. Lin. Soc. Lond*. 23:495–515.

Bateson, William, 1894. *Materials for the Study of Variation*. London: Macmillan.

———, 1902. *Mendel's Principles of Heredity*. Cambridge: Cambridge University Press.

Beatty, John, 1982. "What's in a Word? Coming to Terms in the Darwinian Revolution." *J. Hist. Biol*. 15:214–39.

———, 1985. "Speaking of Species: Darwin's Strategy." In *The Darwinian Heritage*, ed David Kohn, 265–82. Princeton: Princeton University Press.

Beddall, Barbara G., 1969. *Wallace and Bates in the Tropics*. London: Macmillan.

Beer, Gillian, 1983. *Darwin's Plots: Evolutionary Narrative in Darwin, George Eliot, and Nineteenth-Century Fiction*. London: Routledge and Kegan Paul.

Bellomy, Donald C., 1984. " 'Social Darwinism' Revisited." *Perspectives in American History*, n.s. 1:1–129.

Bennett, J. H., ed., 1983. *Natural Selection, Heredity, and Eugenics*. Oxford: Clarendon Press.

Benson, K. R., 1981. "Problems of Individual Development: Descriptive Embryology in America at the Turn of the Century." *J. Hist. Biol.* 14:115–28.

Bernal, J. D., 1969. *Science in History*. 4 vols. Harmondsworth, Eng.: Penguin Books.

Blacker, C., 1952. *Eugenics: Galton and After*. Cambridge, Mass.: Harvard University Press.

Boakes, Robert, 1984. *From Darwin to Behaviorism: Psychology and the Minds of Animals*. Cambridge: Cambridge University Press.

Boring, Edwin G. [1929] 1950. *A History of Experimental Psychology*. 2d ed. New York: Appleton-Century-Crofts.

Boule, Marcellin, 1909–11. "L'homme fossile de La Chapelle-aux-Saints." *Annales de paleontologie* 6:109–72; 7:18–56, 85–192; 8:1–71.

———, 1921. *Les hommes fossils*. Paris: Masson.

Bourdier, Frank, 1969. "Geoffroy Saint Hilaire versus Cuvier: The Campaign for Paleontological Evolution." In *Toward a History of Geology*, ed. C. J. Schneer, 36–61. Cambridge, Mass.: MIT Press.

Bowler, Peter J., 1974. "Darwin's Concepts of Variation." *J. Hist. Medicine* 29:196–212.

———, 1975. "The Changing Meaning of 'Evolution.'" *J. Hist. Ideas* 36:95–114.

———, 1976a. *Fossils and Progress: Paleontology and the Idea of Progressive Evolution in the Nineteenth Century*. New York: Science History Publications.

———, 1976b. "Malthus, Darwin, and the Concept of Struggle." *J. Hist. Ideas* 37:631–50.

———, 1976c. "Alfred Russel Wallace's Concepts of Variation." *J. Hist. Medicine* 31:17–29.

———, 1977a. "Darwinism and the Argument from Design: Suggestions for a Re-evaluation." *J. Hist. Biol.* 10:29–43.

———, 1977b. "Edward Drinker Cope and the Changing Structure of Evolutionary Theory." *Isis* 68:249–65.

———, 1978. "Hugo De Vries and Thomas Hunt Morgan: The Mutation Theory and the Spirit of Darwinism." *Annals of Sci.* 35:55–73.

———, 1979. "Theodor Eimer and Orthogenesis: Evolution by Definitely Directed Variation." *J. Hist. Medicine* 34:40–73.

———, 1983. *The Eclipse of Darwinism: Anti-Darwinian Evolution Theories in the Decades around 1900*. Baltimore, Johns Hopkins University Press.

———, 1984a. *Evolution: The History of an Idea*. Berkeley and Los Angeles: University of California Press.

———, 1984b. "E. W. MacBride's Lamarckian Eugenics." *Annals of Sci.* 41:245–60.

————, 1985. "Scientific Attitudes to Darwinism in Britain and America." In *The Darwinian Heritage*, ed. David Kohn, 641–81. Princeton: Princeton University Press.

————, 1986. *Theories of Human Evolution: A Century of Debate, 1844–1944.* Baltimore: Johns Hopkins University Press; Oxford: Basil Blackwell.

Box, Joan F., 1978. *R. A. Fisher: The Life of a Scientist.* New York: John Wiley.

Brace, C. Loring, 1982. "The Roots of the Race Concept in American Physical Anthropology." In *A History of American Physical Anthropology*, ed. Frank Spencer, 11–29. New York: Academic Press.

Brackman, Arnold C., 1980. *A Delicate Arrangement: The Strange Case of Charles Darwin and Alfred Russel Wallace.* New York: Times Books.

Brent, Peter, 1981. *Charles Darwin: A Man of Enlarged Curiosity.* New York: Harper and Row.

Brock, W. H., and R. MacLeod, 1976. "The Scientists' Declaration: Reflections on Science and Belief in the Wake of *Essays and Reviews.*." *Brit. J. Hist. Sci.* 9:39–66.

Bronn, H. G., 1858. *Untersuchungen über die Entwickelungsgesetze der organischen Welt.* Stuttgart: E. Schweizerbart.

————, 1859. "On the Laws of Evolution of the Organic World." *Annals and Magazine of Natural History*, 3d ser. 4:81–89.

Brooke, John Hedley, 1985. "The Relations between Darwin's Science and his Religion." In *Darwinism and Divinity*, ed. John Durant, 40–75. Oxford: Basil Blackwell.

Brooks, John L., 1984. *Just before the Origin: Alfred Russel Wallace's Theory of Evolution.* New York: Columbia University Press.

Brown, Frank B., 1986. "The Evolution of Darwin's Theism." *J. Hist. Biology* 19:1–45.

Browne, Janet, 1983. *The Secular Ark: Studies in the History of Biogeography.* New Haven: Yale University Press.

Brush, Stephen, 1987. "The Nebular Hypothesis and the Evolutionary World View." *History of Science* 35:223–43.

Burchfield, Joe D., 1974. "Darwin and the Dilemma of Geological Time." *Isis* 65:301–21.

————, 1975. *Lord Kelvin and the Age of the Earth.* New York: Science History Publications.

Burke, James, 1985. *The Day the Universe Changed.* London: BBC.

Burkhardt, F., 1974. "England and Scotland: The Learned Societies." In *The Comparative Reception of Darwinism*, ed. Thomas F. Glick, 32–74. Austin: University of Texas Press.

Burkhardt, Richard W., Jr., 1977. *The Spirit of System: Lamarck and Evolutionary Biology.* Cambridge, Mass.: Harvard University Press.

————, 1985. "Darwinian Animal Behavior and Evolution." In *The Darwinian Heritage*, ed. David Kohn, 327–66. Princeton: Princeton University Press.

Burrow, J. W., 1966. *Evolution and Society: A Study in Victorian Social Theory.* Cambridge: Cambridge University Press.

Buss, Alan R., 1976. "Galton and the Birth of Differential Psychology and Eugenics." *J. Hist. Behavioral Sci.* 12:47–58.

Butler, Samuel, 1879. *Evolution, Old and New.* London: Harwick and Bogue.

————, 1908. *Essays on Life, Art, and Science.* London: Fifield.

————[1887] 1920. *Luck or Cunning as the Main Means of Organic Modification.* 2d ed. London: Fifield.

Butterfield, Herbert, 1931. *The Whig Interpretation of History.* London: G. Bell and Sons.

Cannon, H. G., 1959. *Lamarck and Modern Genetics.* Springfield, Ill.: Charles C. Thomas.

Cannon, Walter, 1960. The Uniformitarian-Catastrophist Debate." *Isis* 51:38–55.

————, 1961. "The Bases of Darwin's Achievement: A Revolution." *Victorian Studies* 5:109–32.

Caplan, Arthur L., 1978. *The Sociobiology Debate.* New York: Harper and Row.

Carlson, Elof Axel, 1966. *The Gene: A Critical History.* Philadelphia: W. B. Saunders.

Carpenter, W. B., 1888. *Nature and Man: Essays Scientific and Philosophical.* London: Kegan Paul, Trench, Trubner.

Chadwick, Owen, 1975. *The Secularization of the European Mind in the Nineteenth Century.* Cambridge: Cambridge University Press.

Chamberlin, J. Edward, and Sander L. Gilman, eds., 1985. *Degeneration: The Dark Side of Progress.* New York: Columbia University Press.

Chambers, Robert, 1844, 1846, 1860. *Vestiges of the Natural History of Creation.* 1st, 6th, and 11th eds. London: Churchill.

Chapman, Roger, and Cleveland T. Duval, eds., 1982. *Charles Darwin, 1809–1882: A Centennial Commemorative.* Wellington, New Zealand: Nova Pacifica.

Cherfas, Jeremy, 1982. *Darwin Up to Date.* London: IPC Magazines.

Churchill, Frederick B., 1970. "Hertwig, Weismann, and the Meaning of Reduction Division." *Isis* 61:429–57.

————, 1982. "Darwin and the Historian." *Biol. J. Lin. Soc.* 17:45–68.

————, 1986. "Weismann, Hydromedusae, and the Biogenetic Imperative: A Reconsideration." In *A History of Embryology,* ed. T. J. Horder, J. A. Witkowski, and C. C. Wylie, 7–34. Cambridge: Cambridge University Press.

Clark, Linda L., 1984. *Social Darwinism in France.* University, Ala.: University of Alabama Press.

Clark, Ronald, 1984. *The Survival of Charles Darwin.* London: Weidenfeld and Nicholson.

Clark, W. E. Le Gros, 1934. *Early Forerunners of Man.* London: Ballière, Tyndall and Cox.

———, 1955. *The Fossil Evidence for Human Evolution*. Chicago: University of Chicago Press.

———, 1967. *Man-Apes or Ape-Men?* New York: Holt, Rinehart and Winston.

Cock, A. G., 1973. "William Bateson, Mendelism, and Biometry." *J. Hist. Biol.* 6:1–36.

Cohen, I. Bernard, 1985. *Revolution in Science*. Cambridge, Mass.: Harvard University Press.

Coleman, William, 1965. "Cell, Nucleus, and Inheritance: An Historical Study." *Proc. Am. Phil. Soc.* 109:124–58.

———, 1971. *Biology in the Nineteenth Century*. New York: John Wiley.

———, 1976. "Morphology between Type Concept and Descent Theory." *J. Hist. Medicine* 31:149–75.

Colp, Ralph, Jr., 1974. "The Contacts between Karl Marx and Charles Darwin." *J. Hist. Ideas* 35:329–38.

———, 1986. "Confessing a Murder: Darwin's First Revelations about Transmutation." *Isis* 77:9–32.

Conry, Yvette, 1974. *L'introduction du Darwinisme en France au XIXe siècle*. Paris: J. Vrin.

Cooter, Roger, 1985. *The Cultural Meaning of Popular Science: Phrenology and the Organization of Consent in Nineteenth-Century Britain*. Cambridge: Cambridge University Press.

Cope, Edward Drinker, 1868. "On the Origin of Genera." *Proc. Acad. Nat. Sci. Philadelphia* 20:242–300.

———, 1887a. *The Origin of the Fittest*. New York: Macmillan.

———, 1887b. *Theology of Evolution: A Lecture*. Philadelphia: Arnold.

Corsi, Pietro, 1978. "The Importance of French Transformist Ideas for the Second Volume of Lyell's *Principles of Geology*." *Brit. J. Hist. Sci.* 11:221–44.

Corsi, Pietro, and Paul Weindling, 1985. "Darwinism in Germany." In *The Darwinian Heritage*, ed. David Kohn, 683–729. Princeton: Princeton University Press.

Cowan, Ruth Schwartz, 1972a. "Francis Galton's Contributions to Genetics." *J. Hist. Biol.* 5:389–412.

———, 1972b. "Francis Galton's Statistical Ideas: The Influence of Eugenics." *Isis* 63:509–28.

———, 1977. "Nature and Nurture: The Interplay of Biology and Politics in the Work of Francis Galton." *Stud. Hist. Biol.* 1:133–208.

Cravens, Hamilton, 1978. *The Triumph of Evolution: American Scientists and the Heredity-Environment Controversy, 1900–1941*. Philadelphia: University of Pennsylvania Press.

Crook, D. P., 1984. *Benjamin Kidd: Portrait of a Social Darwinist*. Cambridge: Cambridge University Press.

Daniel, Glyn, 1975. *A Hundred and Fifty Years of Archaeology*. London: Duckworth.

Darlington, C. D., 1959. *Darwin's Place in History*. Oxford: Basil Blackwell.

Darwin, Charles Robert, 1859. *On the Origin of Species by Means of Natural Selection; or, The Preservation of Favoured Races in the Struggle for Life.* London: John Murray.

———, 1871. *The Descent of Man and Selection in Relation to Sex.* 2 vols. London: John Murray.

———, 1887. *The Life and Letters of Charles Darwin,* ed. Francis Darwin. 3 vols. London: John Murray.

———, 1958. *The Autobiography of Charles Darwin, with the Original Omissions Restored,* ed. Nora Barlow. New York: Harcourt, Brace.

———, 1959a. "Darwin's Journal," ed. Sir Gavin De Beer. *Bull. Brit. Mus. (Nat. Hist.), Historical Series* 2, no. 1.

———, 1959b. *The Origin of Species: A Variorum Text,* ed. Morse Peckham. Philadelphia: University of Pennsylvania Press.

———, 1960–61. "Darwin's Notebooks on the Transmutation of Species," ed. Sir Gavin De Beer. *Bull. Brit. Mus. (Nat. Hist.), Historical Series* 2, nos. 2–6.

———, 1963. "Darwin's Ornithological Notes," ed. Nora Barlow. *Bull. Brit. Mus. (Nat. Hist.), Historical Series* 2, no. 7.

———, 1967. "Darwin's Notebooks . . . Pages Excised by Darwin," ed. Sir Gavin De Beer. *Bull. Brit. Mus. (Nat. Hist.), Historical Series* 3, no. 5.

———, 1974. "Notebooks on Man," ed. Paul H. Barrett. In *Darwin on Man,* by Howard E. Gruber. New York: E. P. Dutton.

———, 1975. *Charles Darwin's Natural Selection, Being the Second Part of His Big Species Book Written from 1856 to 1858,* ed. Robert C. Stauffer. London: Cambridge University Press.

———, 1977. *The Collected Papers of Charles Darwin,* ed. Paul H. Barrett. 2 vols. Chicago: University of Chicago Press.

———, 1979. *The Red Notebooks of Charles Darwin,* ed. Sandra Herbert. Ithaca, N.Y.: Cornell University Press.

———, 1985a. *The Correspondence of Charles Darwin.* Vol. 1, *1821–36.* Cambridge: Cambridge University Press.

———, 1985b. *The Formation of Vegetable Mould through the Action of Earthworms.* Introd. Stephen Jay Gould. Chicago: University of Chicago Press.

———, 1986. *The Correspondence of Charles Darwin.* Vol. 2, *1837–42.* Cambridge: Cambridge University Press.

———, 1987. *Charles Darwin's Notebooks, 1836–1844.* Cambridge: Cambridge University Press.

Darwin, Charles, and Alfred Russel Wallace, 1958. *Evolution by Natural Selection.* Cambridge: Cambridge University Press.

Darwin, Erasmus [1794–96] 1974. *Zoonomia; or, the Laws of Organic Life.* Reprint. 2 vols. New York: AMS Press.

Dawkins, W. Boyd, 1874. *Cave Hunting.* London: Macmillan.

De Beer, Sir Gavin, 1963. *Charles Darwin.* London: Nelson.

De Marrais, R., 1974. "The Double-Edged Effect of Sir Francis Galton:

A Search for the Motives of the Biometrician-Mendelian Debate."
J. Hist. Biol. 7:141–74.

de Mortillet, Gabriel, 1883. *Le Préhistorique.* Paris: Reinwald.

Dennert, E., 1904. *At the Deathbed of Darwinism.* Burlington, Iowa: German Literary Board.

Desmond, Adrian, 1982. *Archetypes and Ancestors: Palaeontology in Victorian London, 1850–1875.* London: Blond and Briggs.

———, 1984. "Robert E. Grant: The Social Predicament of a Pre-Darwinian Evolutionist." *J. Hist. Biol.* 17:189–223.

———, 1987, "Artisan Resistance and Evolution in Britain, 1819–1848." *Osiris,* 2d ser. 3:77–110.

———, 1989. *The Politics of Evolution.* Chicago: University of Chicago Press.

De Vries, Hugo, 1910a. *Intracellular Pangenesis.* Chicago: Open Court.

———, 1910b. *The Mutation Theory.* 2 vols. London: Kegan Paul, Trench, Trubner.

Di Gregorio, Mario, 1982. "The Dinosaur Connection: A Reinterpretation of T. H. Huxley's Evolutionary View." *J. Hist. Biol.* 15:397–418.

———, 1984. *T. H. Huxley's Place in Natural Science.* New Haven: Yale University Press.

Dobzhansky, Theodosius, 1937. *Genetics and the Origin of Species.* New York: Columbia University Press.

———, 1962. *Mankind Evolving.* New York: Oxford University Press.

Draper, J. W., 1874. *History of the Conflict between Religion and Science.* New York: Appleton.

Drummond, Henry, 1894. *The Ascent of Man.* New York: James Pott.

Dunn, L. C., 1965. *A Short History of Genetics.* New York: McGraw-Hill.

Dupree, A. Hunter, 1959. *Asa Gray.* Cambridge, Mass.: Harvard University Press.

Durant, John, 1979. "Scientific Naturalism and Social Reform in the Thought of Alfred Russel Wallace." *Brit. J. Hist. Sci.* 12:31–58.

———, 1985a. *Darwinism and Divinity: Essays on Evolution and Religious Belief.* Oxford: Basil Blackwell.

———, 1985b. "Darwinism and Divinity: A Century of Debate." In *Darwinism and Divinity,* ed. Durant, 9–39.

Eimer, G. H. Theodor, 1890. *Organic Evolution as the Result of the Inheritance of Acquired Characters.* London: Macmillan.

———, 1898. *On Orthogenesis and the Impotence of Natural Selection in Species Formation.* Chicago: Open Court.

Eiseley, Loren, 1958. *Darwin's Century: Evolution and the Men Who Discovered It.* New York: Doubleday.

———, 1959. "Charles Darwin, Edward Blyth and the Theory of Natural Selection." *Proc. Am. Phil. Soc.* 103:94–158.

———, 1979. *Darwin and the Mysterious Mr. X.* London: J. M. Dent.

Eldredge, Niles, 1986. *Time Frames: The Rethinking of Darwinian Evolution and the Theory of Punctuated Equilibria.* London: Heinemann.

Ellegård, Alvar, 1958. *Darwin and the General Reader: The Reception of Darwin's Theory of Evolution in the British Periodical Press, 1859–1872.* Göteburg: Acta Universitatis Gothenburgensis.

Ellenberger, Henri, 1970. *The Discovery of the Unconscious.* London: Allen Lane.

Evans, Brian, and Bernard Waites, 1981. *IQ and Mental Testing: An Unnatural Science and Its Social History.* London: Macmillan.

Farley, John, 1974. "The Initial Reaction of French Biologists to Darwin's *Origin of Species.*" *J. Hist. Biol.* 7:275–300.

———, 1982. *Gametes and Spores: Ideas about Sexual Reproduction, 1750–1914.* Baltimore: Johns Hopkins University Press.

Farrall, Lynsay A., 1979. "The History of Eugenics: A Bibliographical Review." *Annals of Sci.* 36:111–23.

Fay, Margaret A., 1978. "Did Marx Offer to Dedicate *Capital* to Darwin?" *J. Hist. Ideas* 39:133–46.

Feuer, Lewis, 1975. "Is the Darwin-Marx Correspondence Authentic?" *Annals of Sci.* 32:1–12.

———, 1976. "On the Darwin-Marx Correspondence." *Annals of Sci.* 33:383–94.

Fichman, Martin, 1977. "Wallace, Zoogeography, and the Problem of Land Bridges." *J. Hist. Biol.* 10:45–63.

———, 1981. *Alfred Russel Wallace.* Boston: Twayne.

———, 1984. "Ideological Factors in the Dissemination of Darwinism." In *Transformation and Tradition in the Sciences,* ed. Everett Mendelsohn, 471–85. Cambridge, Mass.: Harvard University Press.

Fisher, Ronald Aylmer, 1918. "The Correlation between Relatives on the Supposition of Mendelian Inheritance." *Trans. Roy. Soc. Edinburgh* 52:399–433.

———, 1930. *The Genetical Theory of Natural Selection.* Oxford: Clarendon Press.

Fiske, John, 1874. *Outlines of Cosmic Philosophy.* 2 vols. London: Macmillan.

Forrest, D., 1974. *Francis Galton: The Life and Work of a Victorian Genius.* New York: Tapplinger.

Foucault, Michel, 1970. *The Order of Things: The Archaeology of the Human Sciences.* New York: Pantheon.

Frazer, James, 1913. *Psyche's Task: A Discourse Concerning the Influence of Superstition on the Growth of Institutions.* London: Macmillan.

———[1890] 1924. *The Golden Bough: A Study in Magic and Religion.* Abr. ed. London: Macmillan.

Freeland, Guy, 1983. "Evolution and Arch(a)eology." In *The Wider Domain of Evolutionary Thought,* ed. D. Oldroyd and I. Langham, 175–219. Dordrecht: D. Reidel.

Freeman, Derek, 1974. "The Evolutionary Theories of Charles Darwin and Herbert Spencer." *Current Anthropology* 15:211–37.

Freud, Sigmund, 1959–74. *The Standard Edition of the Complete Psychological Works of Sigmund Freud.* 24 vols. London: Hogarth Press.

Friedman, A. J., and C. Donley, 1985. *Einstein as Myth and Muse.* Cambridge: Cambridge University Press.

Froggatt, P., and N. C. Nevin, 1971. "The 'Law of Ancestral Heredity' and the Mendelian-Ancestrian Controversy in England, 1889–1900." *J. Med. Genetics* 8:1–36.

Gale, Barry G., 1972. "Darwin and the Concept of the Struggle for Existence: A Study in the Extra-scientific Origins of Scientific Ideas." *Isis* 63:321–44.

———, 1982. *Evolution without Evidence: Charles Darwin and the Origin of Species.* Albuquerque: University of New Mexico Press.

Galton, Francis, 1869. *Hereditary Genius.* London: Macmillan.

———, 1889. *Natural Inheritance.* London: Macmillan.

———, 1908. *Memories of My Life.* London: Methuen.

Garson, Robert, and Richard Maidment, 1981. "Social Darwinism and the Liberal Tradition: The Case of William Graham Sumner." *South Atlantic Quarterly* 80:61–76.

Gasman, Daniel, 1971. *The Scientific Origins of National Socialism: Social Darwinism in Ernst Haeckel and the Monist League.* New York: American Elsevier.

———, 1985. "Haeckel's Religious Monism: Its Cultural Impact." *Acts XVIIth Internat. Cong. Hist. Sci.,* vol. 1, sess. 16.3.

George, Wilma, 1964. *Biologist-Philosopher: A Study of the Life and Writings of Alfred Russel Wallace.* New York: Abelard-Schumann.

Gilbert, G. N., and M. Mulkay, 1984. *Opening Pandora's Box.* Cambridge: Cambridge University Press.

Gilbert, Scott F., 1978. "The Embryological Origins of the Gene Theory." *J. Hist. Biol.* 11:307–51.

Gillespie, Neal C., 1977. "The Duke of Argyll, Evolutionary Anthropology, and the Art of Scientific Controversy." *Isis* 68:40–54.

Gillispie, Charles C., 1959. *Genesis and Geology.* New York: Harper and Row.

Glass, Bentley, Owsei Temkin, and William L. Strauss, eds., 1959. *Forerunners of Darwin, 1745–1859.* Baltimore: Johns Hopkins Press.

Glick, Thomas F., ed., 1974. *The Comparative Reception of Darwinism.* Austin: University of Texas Press.

Godfrey, Laurie, 1985. "Darwinian, Spencerian, and Modern Perspectives on Progress in Biological Evolution." In *What Darwin Began,* ed. Laurie Godfrey, 40–60. Boston: Allyn and Bacon.

Goldman, Eric, 1952. *Rendezvous with Destiny.* New York: Alfred Knopf.

Gould, Stephan Jay, 1977a. *Ever Since Darwin: Reflections on Natural History.* New York: W. W. Norton.

———, 1977b. *Ontogeny and Phylogeny.* Cambridge, Mass: Harvard University Press.

———, 1980. "G. G. Simpson, Paleontology, and the Modern Synthesis." In *The Evolutionary Synthesis,* ed. Ernst Mayr and William Provine, 152–72. Cambridge, Mass.: Harvard University Press.

———, 1981. *The Mismeasure of Man.* New York: W. W. Norton.

———, 1983. "The Hardening of the Modern Synthesis." In *Dimensions of Darwinism,* ed. Marjorie Grene, 71–93. Cambridge: Cambridge University Press.

———, 1987. "The Fossil Feud that Never Was." *New Scientist* 113:32–37.

Gray, Asa, 1876. *Darwiniana: Essays and Reviews pertaining to Darwinism.* New York: Appleton.

Grayson, Donald K., 1983. *The Establishment of Human Antiquity.* New York: Academic Press.

Greene, John C., 1959. *The Death of Adam: Evolution and Its Impact on Western Thought.* Ames: Iowa State University Press.

———, 1977. "Darwin as a Social Evolutionist." *J. Hist. Biol.* 10:1–27.

———, 1981. *Science, Ideology, and World View.* Berkeley and Los Angeles: University of California Press.

———, 1986. "The History of Ideas Revisited." *Revue de Synthèse* 107:201–28.

Greene, Mott T., 1982. *Geology in the Nineteenth Century.* Ithaca, N.Y.: Cornell University Press.

Greenwood, Davydd J., 1984. *The Taming of Evolution: The Persistence of Nonevolutionary Views in the Study of Humans.* Ithaca, N.Y.: Cornell University Press.

Grehan, J. R., and Ruth Ainsworth, 1985. "Orthogenesis and Evolution." *Systematic Zoology* 34:174–92.

Grene, Marjorie, ed., 1983. *Dimensions of Darwinism: Themes and Counterthemes in Twentieth-Century Evolutionary Theory.* Cambridge: Cambridge University Press.

Gruber, Howard E., 1974. *Darwin on Man: A Psychological Study of Scientific Creativity.* New York: E. P. Dutton.

Gruber, Jacob W., 1960. *A Conscience in Conflict: The Life of St. George Jackson Mivart.* New York: Columbia University Press.

———, 1964. "Brixham Cave and the Antiquity of Man." In *Context and Meaning in Cultural Anthropology,* ed. Milford E. Spiro, 373–402. New York: Free Press.

Haeckel, Ernst, 1866. *Generelle Morphologie der Organismen.* Berlin: Reimer.

———, 1868. *Natürliche Schöpfungsgeschichte: gemeinverstanliche wissenschaftliche Vorträge über die Entwickelungslehre in Allgemeinen und die jenige von Darwin, Goethe, und Lamarck in Besonderen.* Berlin: Reimer.

———, 1874. *Anthropogenie; oder, Entwickelungsgeschichte des Menschen.* Leipzig: Engelmann.

———, 1876a. *The History of Creation; or, the Development of the Earth and its Inhabitants by the Action of Natural Causes; a Popular Exposition of the Doctrine of Evolution in General and of that of Darwin, Goethe, and Lamarck in Particular.* 2 vols. New York: Appleton.

————, 1876b. *Die Perigenesis der Plastidule*. Berlin: Reimer.

————, 1879. *The Evolution of Man*. 2 vols. New York: Appleton.

————, 1900. *The Riddle of the Universe*. London: Watts.

Haldane, J. B. S., 1932. *The Causes of Evolution*. London: Longmans.

————, 1938. *Heredity and Politics*. London: Allen and Unwin.

Haller, John S., 1971. *Outcasts from Evolution: Scientific Attitudes of Racial Inferiority, 1859–1900*. Urbana: University of Illinois Press.

Haller, Mark H., 1963. *Eugenics: Hereditarian Attitudes in American Thought*. New Brunswick, N.J.: Rutgers University Press.

Halliday, R. J., 1971. "Social Darwinism: A Definition." *Victorian Studies* 14:389–405.

Hammond, Michael, 1980. "Anthropology as a Weapon of Social Combat in Late Nineteenth-Century France." *J. Hist. Behavioral Sci.* 16:118–32.

————, 1982. "The Expulsion of the Neanderthals from Human Ancestry." *Social Studies of Science* 12:1–36.

Hardie, J. Keir, 1907. *From Serfdom to Socialism*. London: George Allen.

Hardy, Sir Alistair, 1984. *Darwin and the Spirit of Man*. London: Collins.

Harris, Marvin, 1968. *The Rise of Anthropological Theory*. New York: Thomas Y. Crowell.

Harwood, Jonathan, 1984. "The Reception of Morgan's Chromosome Theory in Germany." *Medizin historisches Journal* 19:3–32.

————, 1985. "Genetics and the Evolutionary Synthesis in Interwar Germany." *Annals of Sci.* 42:279–301.

Hatch, Elvin, 1973. *Theories of Man and Culture*. New York: Columbia University Press.

Helfand, M. S., 1977. "T. H. Huxley's 'Evolutionary Ethics.'" *Victorian Studies* 20:159–77.

Henkin, Leo J., 1963. *Darwinism in the English Novel, 1860–1910*. New York: Russell and Russell

Herschel, Sir J. F. W., 1861. *Physical Geography*. Edinburgh: A. and C. Black.

Heyer, Paul, 1982. *Nature, Human Nature, and Society: Marx, Darwin, Biology, and the Human Sciences*. Westport, Conn.: Greenwood Press.

Himmelfarb, Gertrude, 1959. *Darwin and the Darwinian Revolution*. New York: W. W. Norton.

————, 1968. "Varieties of Social Darwinism." In *Victorian Minds*, 314–32. London: Weidenfeld and Nicholson.

Hodge, M. J. S., 1971. "Lamarck's Science of Living Bodies." *Brit. J. Hist. Sci.* 5:323–52.

————, 1972. "The Universal Gestation of Nature: Chambers' *Vestiges* and *Explanations*." *J. Hist. Biol.* 5:127–52.

————, 1982. "Darwin and the Laws of the Animate Part of the Terrestrial System (1835–1837)." *Stud. Hist. Biol.* 7:1–106.

————, 1985a. "Darwin as a Lifelong Generation Theorist." In *The Darwinian Heritage*, ed. David Kohn, 207–44. Princeton: Princeton University Press.

————, 1985b. "Generation and the Origin of Species from Darwin (1837)

to Dobzhansky (1937): Historiographic Proposals." *Acts XVIIth International. Cong. Hist. Sci.*, vol. 2, sess. 3.2.

Hodge, M. J. S., and David Kohn, 1985. "The Immediate Origins of Natural Selection." In *The Darwinian Heritage*, ed. Kohn, 185–206. Princeton: Princeton University Press.

Hofstadter, Richard [1944] 1955. *Social Darwinism in American Thought.* Rev. ed. Boston: Beacon Press.

Hooker, Joseph Dalton, 1860. "On the Origination and Distribution of Vegetable Species." *Am. J. Sci.*, 2d ser. 29:1–25, 305–26.

Hooykaas, R., 1959. *Natural Law and Divine Miracle*. Leiden: Brill.

———, 1966. "Geological Uniformitarianism and Evolution." *Archives Internat. Hist. Sci.* 19:3–19.

Horder, T. J., J. A. Witkowski, and C. C. Wylie, eds., 1986. *A History of Embryology*. Cambridge: Cambridge University Press.

Hoyle, Fred, and C. Wickramsinghe, 1986. *Archaeopteryx: The Primordial Bird*. Swansea, Wales: Christopher Davies.

Hull, David L., 1973. *Darwin and His Critics*. Cambridge, Mass.: Harvard University Press.

———, 1978. "Sociobiology: A Scientific Bandwaggon or a Travelling Medicine Show?" In *Sociobiology and Human Nature*, ed. M. S. Gregory, 136–63. San Francisco: Jossey-Bass.

———, 1985. "Darwinism as a Historical Entity: A Historiographical Proposal." In *The Darwinian Heritage*, ed. David Kohn, 773–812. Princeton: Princeton University Press.

Hull, David L., Peter D. Tessner, and Arthur M. Diamond, 1978. "Planck's Principle: Do Younger Scientists Accept New Ideas with Greater Alacrity than Older Scientists?" *Science* 202:717–23.

Huxley, Julian S., 1942. *Evolution: The Modern Synthesis*. London: Allen and Unwin.

———, 1957. *New Bottles for New Wine*. New York: Harper and Row.

Huxley, Thomas Henry, 1854. "Vestiges of the Natural History of Creation." *Brit. & Foreign Med. Chirug. Rev.* 13:332–43.

———, 1863. *Man's Place in Nature*. London: Williams and Norgate.

———, 1877. *American Addresses*. London: Macmillan.

———, 1893–94. *Collected Essays*. Vol. 2, *Darwiniana*; vol. 4, *Science and Hebrew Tradition*; vol. 7, *Man's Place in Nature*; vol. 9, *Evolution and Ethics*. London: Macmillan.

Hyatt, Alpheus, 1866. "On the Parallelism between the Different Stages of Life in the Individual and Those in the Entire Group of the Molluscous Order Tetrabranchiata." *Mem. Boston Soc. Nat. Hist.* 1:193–209.

———, 1884. "Evolution of the Cephalopoda." *Science* 3:122–27.

———, 1889. *Genesis of the Arietidae*. Smithsonian Contributions to Knowledge, no. 673. Washington, D.C.: Smithsonian Institution.

Hyman, Stanley, 1962. *The Tangled Bank: Darwin, Marx, Frazer, and Freud as Imaginative Writers*. New York: Atheneum.

Jenkin, Fleeming, 1867. "The Origin of Species." *North British Review* 46:277–318.

Jones, Greta, 1980. *Social Darwinism in English Thought*. London: Harvester.

———, 1986. *Social Hygiene in Twentieth-Century Britain*. London: Croom Helm.

Jones, Henry Festing, 1919. *Samuel Butler: A Memoir*. 2 vols. London: Macmillan.

Joravsky, D., 1970. *The Lysenko Affair*. Cambridge, Mass.: Harvard University Press.

Jowett, B. et al., 1860. *Essays and Reviews*. London: John Parker.

Judd, J. W., 1911. *The Coming of Evolution: The Story of a Great Revolution in Science*. Cambridge: Cambridge University Press.

Kammerer, Paul, 1923. "Breeding Experiments on the Inheritance of Acquired Characters." *Nature* 111:637–40.

———, 1924. *The Inheritance of Acquired Characteristics*. New York: Boni and Liveright.

Kaye, Howard L., 1986. *The Social Meaning of Modern Biology*. New Haven: Yale University Press.

Keith, Arthur, 1915. *The Antiquity of Man*. London: Williams and Norgate.

———, 1934. *The Construction of Man's Family Tree*. London: Watts.

———, 1948. *A New Theory of Human Evolution*. London: Watts.

Kellogg, Vernon L., 1907. *Darwinism Today: A Discussion of Present-Day Scientific Criticism of the Darwinian Selection Theories*. New York: Henry Holt.

Kelly, Alfred, 1981. *The Descent of Darwin: The Popularization of Darwinism in Germany, 1860–1914*. Chapel Hill: University of North Carolina Press.

Kelvin, William Thomson, baron, 1894. *Popular Lectures and Addresses*. Vol. 2, *Geology and General Physics*. London: Macmillan.

Kennedy, James G., 1978. *Herbert Spencer*. Boston: Twayne.

Kevles, Daniel J., 1985. *In the Name of Eugenics: Genetics and the Uses of Human Heredity*. New York: Knopf.

Keynes, John Maynard, 1971–79. *The Collected Writings of John Maynard Keynes*. 29 vols. London: Macmillan.

Kidd, Benjamin, 1894. *Social Evolution*. New York: Macmillan.

Kingsley, Charles [1876] 1877. *Charles Kingsley: His Letters and Memories of his Life*. 2 vols. 4th ed. London: Henry S. King.

———, 1890. *Scientific Lectures and Essays*. New ed. London: Methuen.

Knox, Robert, 1862. *The Races of Man*. 2d ed. London: Renshaw.

Koestler, Arthur, 1967. *The Ghost in the Machine*. London: Macmillan.

———, 1971. *The Case of the Midwife Toad*. London: Hutchinson.

Kohn, David, 1980. "Theories to Work by: Rejected Theories, Reproduction, and Darwin's Path to Natural Selection." *Stud. Hist. Biol.* 4:67–170.

———, 1981. "On the Principle of Diversity." *Science* 213:1105–8.

———, ed., 1985. *The Darwinian Heritage*. Princeton: Princeton University Press.

Kottler, Malcolm, 1974. "Alfred Russel Wallace, the Origin of Man, and Spiritualism." *Isis* 65:145–92.

———, 1985. "Charles Darwin and Alfred Russel Wallace: Two Decades of Debate over Natural Selection." In *The Darwinian Heritage*, ed. David Kohn, 367–432. Princeton: Princeton University Press.

Kropotkin, Peter, 1902. *Mutual Aid: A Factor in Evolution*. London: Heinemann.

———, 1912. "Inheritance of Acquired Characters." *Nineteenth Century and After* 71:511–31.

Kuhn, Thomas S., 1957. *The Copernican Revolution*. Cambridge, Mass.: Harvard University Press.

———, [1962] 1969. *The Structure of Scientific Revolutions*. Reprint. Chicago: University of Chicago Press.

Kuper, A., 1985. "Development of L. H. Morgan's Evolutionism." *J. Hist. Behavioral Sci.* 21:3–22.

Lack, David, 1947. *Darwin's Finches*. Cambridge: Cambridge University Press.

Lamarck, J. B. P. A. [1914] 1963. *Zoological Philosophy*. Trans. Hugh Elliot. London: Macmillan. Reprint. New York: Hafner.

Landau, Misia, 1984. "Human Evolution as Narrative." *Am. Scientist* 72:262–68.

Langham, Ian, 1981. *The Building of British Social Anthropology*. Dordrecht: D. Reidel.

Lanham, Url, 1973. *The Bone Hunters*. New York: Columbia University Press.

Lankester, E. Ray, 1877. "Notes on the Embryology and Classification of the Animal Kingdom." *Quart. J. Microscopical Sci.* 17:399–454.

———, 1880. *Degeneration: A Chapter in Darwinism*. London: Macmillan.

———, 1888. "Zoology." *Encyclopaedia Britannica*, 9th ed. 24:799–810.

Laudan, Rachel, 1987. *From Mineralogy to Geology: The Foundations of a Science, 1650–1830*. Chicago: University of Chicago Press.

La Vergata, A., 1985. "Images of Darwin: A Historiographical Overview." In *The Darwinian Heritage*, ed. David Kohn, 901–72. Princeton: Princeton University Press.

Leatherdale, William, 1983. "The Influence of Darwinism on English Literature." In *The Wider Domain of Evolutionary Thought*, ed. David Oldroyd and Ian Langham, 1–26. Dordrecht: D. Reidel.

Lenoir, Timothy, 1982. *The Strategy of Life: Teleology and Mechanism in Nineteenth-Century German Biology*. Dordrecht: D. Reidel.

Lesch, John E., 1975. "The Role of Isolation in Evolution: George J. Romanes and John T. Gulick." *Isis* 66:483–503.

Lewontin, Richard, 1983. "The Organism as the Subject and as the Object of Evolution." *Scientia* 118:65–80.

Livingstone, David, 1987. *Darwin's Forgotten Defenders: The Encounter between Evangelical Theology and Evolutionary Thought.* Edinburgh: Scottish Universities Press; Grand Rapids, Mich.: Eerdmans.

Lorenz, Konrad, 1966. *On Aggression.* New York: Harcourt, Brace & World.

Lovejoy, Arthur O. [1936] 1960. *The Great Chain of Being: A Study in the History of an Idea.* Reprint. New York: Harper and Row.

Løvtrup, S., 1987. *Darwinism: The Refutation of a Myth.* London: Croom Helm.

Lubbock, John, 1865. *Prehistoric Times.* London: Williams and Norgate.

———, 1870. *On the Origin of Civilisation.* London: Longmans Green.

Lucas, J. R., 1979. "Wilberforce and Huxley: A Legendary Encounter." *Historical Journal* 22:313–30.

Ludmerer, Kenneth, 1972. *Genetics and American Society.* Baltimore: Johns Hopkins University Press.

Lurie, Edward, 1960. *Louis Agassiz: A Life in Science.* Chicago: University of Chicago Press.

Lyell, Charles, 1830–33. *Principles of Geology.* 3 vols. London: John Murray.

———, 1863. *Geological Evidences of the Antiquity of Man.* London: John Murray.

MacBride, E. W., 1914. *Textbook of Embryology.* Vol. 1, *Invertebrates.* London: Macmillan.

MacDougall, William, 1908. *An Introduction to Social Psychology.* London: Methuen.

———, 1927. "An Experiment for the Testing of the Hypothesis of Lamarck." *Brit. J. Psychology* 17:267–304.

Mackenzie, Donald, 1982. *Statistics in Britain, 1865–1930.* Edinburgh: Edinburgh University Press.

McKinney, H. Lewis, ed., 1971. *Lamarck to Darwin: Contributions to Evolutionary Biology.* Lawrence, Kans.: Coronado Press.

———, 1972. *Wallace and Natural Selection.* New Haven: Yale University Press.

MacLeod, Roy, 1969. "The Genesis of *Nature.*" *Nature* 224:423–41.

———, 1970. "The X-Club: A Scientific Network in Late Victorian England." *Notes & Records Roy. Soc. Lond.* 24:305–22.

Maienschein, Jane, 1978. "Cell Lineage, Ancestral Reminiscence, and the Biogenetic Law." *J. Hist. Biol.* 11:129–58.

———, 1981. "Shifting Assumptions in American Biology: Embryology." *J. Hist. Biol.* 14:89–113.

———, 1984. "What Determines Sex? A Study of Converging Research Approaches." *Isis* 75:457–80.

———, 1986. "Preformation or New Formation — or Neither or Both." In *A History of Embryology,* ed. T. J. Horder, J. A. Witkowski, and C. C. Wylie, 73–108. Cambridge: Cambridge University Press.

Maine, Henry, 1861. *Ancient Law.* London: John Murray.

Malthus, Thomas R. [1872] 1914. *An Essay on the Principle of Population.* 7th ed. 2 vols. Reprint. London: Everyman.

————, 1959. *Population: The First Essay.* Ann Arbor: University of Michigan Press.

Mandelbaum, Maurice, 1971. *History, Man, & Reason: A Study in Nineteenth-Century Thought.* Baltimore: Johns Hopkins Press.

Manier, Edward, 1978. *The Young Darwin and His Cultural Circle.* Dordrecht: D. Reidel.

————, 1980. "History, Philosophy, and Sociology of Biology: A Family Romance." *Stud. Hist. & Phil. Sci.* 11:1–24.

Marsh, Othniel C., 1878. *Introduction and Succession of Vertebrate Life in North America.* New York: Appleton.

Marx, Karl, and Friedrich Engels, 1953. *Marx and Engels on Malthus,* ed. Ronald L. Meek. London: Lawrence and Wishart.

————, 1965. *Selected Correspondence.* Moscow: Progress.

Mayr, Ernst, 1942. *Systematics and the Origin of Species.* New York: Columbia University Press.

————, 1954. "Wallace's Line in the Light of Recent Zoogeographic Studies." *Quart. Rev. Biol.* 29:1–14.

————, 1955. "Karl Jordan's Contributions to Current Concepts in Systematics and Evolution." *Trans. Roy. Entomological Soc. London* 107:65–66.

————, 1959a. "Agassiz, Darwin, and Evolution." *Harvard Library Bulletin* 12:165–94.

————, 1959b. "Isolation as an Evolutionary Factor." *Proc. Am. Phil. Soc.* 103:221–30.

————, 1959c. "Where are We?" *Cold Spring Harbor Symposia on Quantitative Biology* 24:409–40.

————, 1964. Introduction to *On the Origin of Species* by Charles Darwin. Cambridge, Mass.: Harvard University Press.

————, 1972a. "The Nature of the Darwinian Revolution." *Science* 176:981–89.

————, 1972b. "Lamarck Revisited." *J. Hist. Biol.* 5:55–94.

————, 1976. *Evolution and the Diversity of Life.* Cambridge, Mass.: Harvard University Press.

————, 1982. *The Growth of Biological Thought: Diversity, Evolution, and Inheritance.* Cambridge, Mass.: Harvard University Press.

————, 1985. "Weismann and Evolution." *J. Hist. Biol.* 18:295–329.

————, 1986. "The Death of Darwin?" *Revue de Synthèse* 107:229–36.

Mayr, Ernst, and W. B. Provine, eds., 1980. *The Evolutionary Synthesis: Perspectives on the Unification of Biology.* Cambridge, Mass.: Harvard University Press.

Medvedev, Z., 1969. *The Rise and Fall of T. D. Lysenko.* New York: Columbia University Press.

Meijer, Onno, 1985. "Hugo De Vries No Mendelian?" *Annals of Science* 42:189–232.

Miller, Hugh [1847] 1851. *Footprints of the Creator: or the Asterolepis of Stromness*. 3d ed. London: Johnstone and Hunter.

Millhauser, Milton, 1959. *Just before Darwin: Robert Chambers and "Vestiges"*. Middletown, Conn.: Wesleyan University Press.

Mivart, St. George Jackson, 1871. *The Genesis of Species*. New York: Appleton.

———, 1884. "On the Development of the Individual and of the Species." *Proc. Zool. Soc. Lond.* 462–74.

———, 1887–88. "On the Possible Dual Origin of the Mammalia." *Proc. Roy. Soc. Lond.* 43:372–79.

Montgomery, William, 1974. "Germany." In *The Comparative Reception of Darwinism*, ed. Thomas F. Glick, 81–116. Austin: University of Texas Press.

Monypenny, William F., and George E. Buckle, [1910–20] 1929. *The Life of Benjamin Disraeli*. Rev. ed. 2 vols. London: John Murray.

Moore, James R., 1979. *The Post-Darwinian Controversies: A Study of the Protestant Struggle to Come to Terms with Darwin in Great Britain and America, 1870–1900*. New York: Cambridge University Press.

———, 1982a. "Charles Darwin Lies in Westminster Abbey." *Biol. J. Lin. Soc.* 17:97–113.

———, 1982b. "1859 and All That: Remaking the Study of Evolution and Religion." In *Charles Darwin: A Centennial Commemorative*, ed. Roger Chapman and C. T. Duval, 167–94. Wellington, New Zealand: Nova Pacifica.

———, 1985a. "Herbert Spencer's Henchmen: The Evolution of Protestant Liberals in Late Nineteenth-Century America." In *Darwinism and Divinity*, ed. John Durant, 76–100. Oxford: Basil Blackwell.

———, 1985b. "Evangelicals and Evolution: Henry Drummond, Herbert Spencer, and the Naturalization of the Spiritual World." *Scottish Journal of Theology* 38:383–417.

———, 1986a. "Socializing Darwinism: Historiography and the Fortunes of a Phase." In *Science as Politics*, ed. Les Levidow, 38–80. London: Free Association Books.

———, 1986b. "Crisis without Revolution: The Ideological Watershed in Victorian England." *Revue de Synthèse*, 4th ser., nos. 1–2, 53–78.

———, ed., in press. *History, Humanity, and Evolution: Essays in Honour of John C. Greene*. New York: Cambridge University Press.

Moorehead, Alan, 1969. *Darwin and the Beagle*. London: Hamish Hamilton.

Morgan, Lewis H. [1877] 1964. *Ancient Society*. Reprint. Cambridge, Mass.: Harvard University Press.

Morgan, Thomas Hunt, 1903. *Evolution and Adaptation*. New York: Macmillan.

———, 1916. *A Critique of the Theory of Evolution*. Princeton: Princeton University Press.

Morgan, T. H., A. H. Sturtevant, H. Muller, and C. B. Bridges, 1915. *The Mechanism of Mendelian Heredity*. New York: Henry Holt.

Morton, Peter, 1984. *The Vital Science: Biology and the Literary Imagination, 1860–1900*. London: Allen and Unwin.

Mulkay, Michael, 1979. *Science and the Sociology of Knowledge*. London: Allen and Unwin.

Munroe, Robert [1893] 1897. *Prehistoric Problems*. Reprint. Edinburgh and London: William Blackwood.

Muschinske, D., 1977. "The Nonwhite as Child: G. Stanley Hall and the Education of Nonwhite Peoples." *J. Hist. Behavioral Sci.* 13:328–36.

Nicholson, A. J., 1960. "The Role of Population Dynamics in Natural Selection." In *Evolution after Darwin*, ed. Sol Tax, 1:477–522. Chicago: University of Chicago Press.

Nisbet, Robert A., 1969. *Social Change and History: Aspects of the Western Theory of Development*. New York: Oxford University Press.

Nordau, Max, 1895. *Degeneration*. London: Heinemann.

Nordenskiöld, Erik [1929] 1946. *The History of Biology*. Reprint. New York: Tudor Publishing.

Norton, B. J., 1973. "The Biometric Defense of Darwinism." *J. Hist. Biol.* 6:283–316.

Numbers, Ronald, 1977. *Creation by Natural Law: Laplace's Nebular Hypothesis in American Thought*. Seattle: University of Washington Press.

Nye, Robert A., 1986. "The Influence of Evolutionary Theory on Pareto's Sociology." *J. Hist. Behavioral Sci.* 22:99–106.

Olby, Robert C., 1979. "Mendel no Mendelian?" *History of Science* 17:53–72.

———[1966] 1985. *The Origins of Mendelism*. 2d ed. Chicago: University of Chicago Press.

Oldroyd, David R., 1984. "How did Darwin Arrive at his Theory?" *History of Science* 22:325–74.

Oldroyd, David, and Ian Langham, eds., 1983. *The Wider Domain of Evolutionary Thought*. Dordrecht: D. Reidel.

Osborn, Henry F., 1894. *From the Greeks to Darwin*. New York: Columbia University Press.

———, 1934. "Aristogenesis: The Creative Principle in the Origin of Species." *Am. Naturalist* 68:193–235.

Ospovat, Dov, 1976. "The Influence of Karl Ernst von Baer's Embryology, 1828–1859." *J. Hist. Biol.* 9:1–28.

———, 1981. *The Development of Darwin's Theory*. Cambridge: Cambridge University Press.

Outram, Dorinda, 1984. *Georges Cuvier: Vocation, Science, and Authority in Post-Revolutionary France*. Manchester: Manchester University Press.

Owen, Richard, 1849. *On the Nature of Limbs*. London: van Voorst.

———, 1860. "Darwin on the Origin of Species." *Edinburgh Rev.* 111:487–532.

———, 1866–68. *On the Anatomy of the Vertebrates*. 3 vols. London: Longmans, Green.

Paley, William, 1802. *Natural Theology; or, Evidences of the Existence and Attributes of the Deity Collected from the Appearances of Nature.* London.

Paradis, Michael, 1978. *T. H. Huxley: Man's Place in Nature.* Lincoln: University of Nebraska Press.

Pastore, Nicholas, 1949. *The Nature-Nurture Controversy.* New York: King's Crown Press.

Pearson, Karl, 1894. "Socialism and Natural Selection." *Fortnightly Rev.,* n.s. 56:1–21.

———, 1896. "Regression, Heredity, and Panmixia." *Phil. Trans. Roy. Soc. Lond.* 197A:253–318.

———, 1898. "Mathematical Contributions to the Theory of Evolution: on the Law of Ancestral Heredity." *Proc. Roy. Soc. Lond.* 57:386–412.

———[1892] 1900. *The Grammar of Science.* 2d ed. London: A. & C. Black.

———, 1901. *National Life from the Standpoint of Science.* London: A. & C. Black.

Peckham, Morse [1959] 1970. "Darwinism and Darwinisticism." Reprinted in Peckham, *The Triumph of Romanticism,* 176–201. Columbia: University of South Carolina Press.

Peel, J. D. Y., 1971. *Herbert Spencer: The Evolution of a Sociologist.* London: Heinemann.

Pfeifer, Edward J., 1965. "The Genesis of American Neo-Lamarckism." *Isis* 56:156–67.

———, 1974. "United States." In *The Comparative Reception of Darwinism.* ed. Thomas F. Glick, 168–206. Austin: University of Texas Press.

Philmus, Robert M., and David Y. Hughes, eds., 1975. *H. G. Wells: Early Writings in Science and Science Fiction.* Berkeley and Los Angeles: University of California Press.

Piaget, Jean, 1979. *Behaviour and Evolution.* London: Routledge and Kegan Paul.

Pickens, D. K., 1968. *Eugenics and the Progressives.* Nashville, Tenn.: Vanderbilt University Press.

Plate, Robert, 1964. *The Dinosaur Hunters: Othniel C. Marsh and Edward D. Cope.* New York: McKay.

Poliakov, L., 1970. *The Aryan Myth.* New York: Basic Books.

Popper, Karl, 1957. *The Poverty of Historicism.* London: Routledge and Kegan Paul.

Poulton, E. B., 1890. *The Colours of Animals.* New York: Appleton.

———, 1908. *Essays in Evolution, 1889–1907.* Oxford: Oxford University Press.

Powell, Baden, 1855. *Essays on the Spirit of the Inductive Philosophy.* London: Longmans.

Provine, William B., 1971. *The Origin of Theoretical Population Genetics.* Chicago: University of Chicago Press.

———, 1978. "The Role of Mathematical Population Genetics in the

Evolutionary Synthesis of the 1930s and 1940s." *Stud. Hist. Biol.* 2:167–92.

———, 1986. *Sewall Wright and Evolutionary Biology.* Chicago: University of Chicago Press.

Radl, Emmanuel, 1930. *The History of Biological Theories.* Oxford: Oxford University Press.

Rainger, Ronald, 1981. "The Continuation of the Morphological Tradition in American Paleontology." *J. Hist. Biol.* 14:129–58.

———, 1985. "Paleontology and Philosophy: A Critique." *J. Hist. Biol.* 18:267–88.

———, 1986. "Just before Simpson: William Diller Matthew's Understanding of Evolution." *Proc. Am. Phil. Soc.* 130:453–74.

Ralling, Christopher, 1978. *The Voyage of Charles Darwin.* London: BBC.

Ratzel, Friedrich, 1896. *The History of Mankind.* 3 vols. London: Macmillan.

Reader, John, 1981. *Missing Links: The Hunt for Earliest Man.* London: Collins.

Rehbock, Philip F., 1983. *The Philosophical Naturalists: Themes in Early Nineteenth-Century British Biology.* Madison: University of Wisconsin Press.

Reif, Wolf-Ernst, 1983. "Evolutionary Theories in German Paleontology." In *Dimensions of Darwinism,* ed. Marjorie Grene, 173–204. Cambridge: Cambridge University Press.

———, 1985. "The Search for a Macroevolutionary Theory in German Paleontology." *J. Hist. Biol.* 19:79–130.

Rensch, Bernhard, 1980. "Historical Development of the Present Synthetic Neo-Darwinism in Germany." In *The Evolutionary Synthesis,* ed. Ernst Mayr and W. B. Provine, 284–303. Cambridge, Mass.: Harvard University Press.

Richards, Evelleen, 1987. "A Question of Property Rights: Richard Owen's Evolutionism Reassessed." *Brit. J. Hist. Sci.* 20:129–72.

Ridley, Mark, 1986. "Embryology and Classical Zoology in Britain." In *A History of Embryology,* ed. T. J. Horder, J. A. Witkowski, and C. C. Wylie, 35–68. Cambridge: Cambridge University Press.

Rivers, W. H. R., 1911. "President's Address, Anthropology Section." *Report of the British Association for the Advancement of Science,* 1911 meeting, 490–99.

Robinson, Gloria, 1979. *A Prelude to Genetics: Theories of a Material Substance of Heredity, Darwin to Weismann.* Lawrence, Kans.: Coronado Press.

Robson, G. C., and O. W. Richards, 1936. *The Variation of Animals in Nature.* London: Longmans, Green.

Rogers, James A., 1972. "Darwinism and Social Darwinism." *J. Hist. Ideas* 33:265–80.

Romanes, G. J., 1888. *Mental Evolution in Man.* London: Kegan Paul, Trench, Trubner.

————, 1892–97. *Darwin and after Darwin.* 3 vols. London: Longmans, Green.

Rotman, Brian, 1977. *Jean Piaget: Psychologist of the Real.* Hassocks, Eng: Harvester.

Rudwick, M. J. S., 1971. "Uniformity and Progression: Reflections on the Structure of Geological Theory in the Age of Lyell." In *Perspectives in the History of Science and Technology,* ed. Duane H. D. Roller, 209–27. Norman: University of Oklahoma Press.

————, 1972. *The Meaning of Fossils: Episodes in the History of Paleontology.* New York: American Elsevier.

————, 1985. *The Great Devonian Controversy.* Chicago: University of Chicago Press.

Ruse, Michael, 1975a. "Charles Darwin and Artificial Selection." *J. Hist. Ideas* 36:339–50.

————, 1975b. "Darwin's Debt to Philosophy." *Stud. Hist. & Phil. of Sci.* 6:159–81.

————, 1979. *The Darwinian Revolution: Science Red in Tooth and Claw.* Chicago: University of Chicago Press.

————, 1982. *Darwinism Defended.* Reading, Mass.: Addison Wesley.

————, 1986. *Taking Darwin Seriously: A Naturalistic Approach to Philosophy.* Oxford: Basil Blackwell.

Russell, E. S., 1916. *Form and Function: A Contribution to the History of Animal Morphology.* London: John Murray.

Russett, Cynthia E., 1976. *Darwin in America: The Intellectual Response.* San Francisco: W. H. Freeman.

Sandler, Iris, and Laurence Sandler, 1985. "A Conceptual Ambiguity that Contributed to the Neglect of Mendel's Paper." *Hist. & Phil. of Life Sci.* 7:3–70.

Santurri, E. N., 1982. "Theodicy and Social Policy in Malthus' Thought." *J. Hist. Ideas* 43:315–30.

Sapp, Jan, 1983. "The Struggle for Authority in the Field of Genetics." *J. Hist. Biol.* 16:311–42.

————, 1987. *Beyond the Gene: Cytoplasmic Inheritance and the Struggle for Authority in Genetics.* New York: Oxford University Press.

Schaffer, Simon, 1986. "Scientific Discoveries and the End of Natural Philosophy." *Social Studies of Science* 16:387–420.

Schwalbe, Gustav, 1904. *Die Vorgeschichte des Menschen.* Braunschweig: Friedrich Vierwig.

————, 1906. *Studien zur Vorgeschichte des Menschen.* Stuttgart: E. Schweizerbartsche.

————, 1909. "The Descent of Man." In *Darwin and Modern Science,* ed. A. C. Seward, 112–36. Cambridge: Cambridge University Press.

Schwartz, Joel S., 1974. "Charles Darwin's Debt to Malthus and Edward Blyth." *J. Hist. Biol.* 7:301–18.

Schweber, Sylvan S., 1977. "The Origin of the *Origin* Revisited." *J. Hist. Biol.* 10:229–326.

Scott, C. H., 1976. *Lester Frank Ward*. Boston: Twayne.

Searle, G. R., 1976. *Eugenics and Politics in Britain, 1900–1914*. Leiden: Noordhoff International.

———, 1979. "Eugenics and Politics in Britain in the 1930s." *Annals of Science* 36:159–69.

Secord, James A., 1986. *Controversy in Victorian Geology*. Princeton: Princeton University Press.

———, in press. "Behind the Veil: The Genesis of Robert Chambers' *Vestiges of the Natural History of Creation*." In *History, Humanity, and Evolution*, ed. James R. Moore. New York: Cambridge University Press.

Semmel, Bernard, 1960. *Imperialism and Social Reform*. London: Allen and Unwin.

Seward, A. C., ed., 1909. *Darwin and Modern Science*. Cambridge: Cambridge University Press.

Shapin, Stephen, 1982. "The History of Science and its Sociological Reconstruction." *History of Science* 20:157–211.

Shaw, George Bernard, 1970–74. *The Bodley Head Bernard Shaw*. 7 vols. London: Bodley Head.

Shor, Elizabeth, 1974. *The Fossil Feud between E. D. Cope and O. C. Marsh*. New York: Exposition Press.

Simpson, George Gaylord, 1944. *Tempo and Mode in Evolution*. New York: Columbia University Press.

———, 1967. *The Meaning of Evolution*. New Haven: Yale University Press.

———, 1973. "The Concept of Progress in Organic Evolution." *Social Research* 41:28–51.

Skinner, B. F., 1948. *Walden Two*. London: Macmillan.

Sloan, Philip R., 1985. "Darwin's Invertebrate Program, 1826–36." In *The Darwinian Heritage*, ed. David Kohn 71–120. Princeton: Princeton University Press.

———, 1986. "Darwin, Vital Matter, and the Transformism of Species." *J. Hist. Biol.* 19:369–445.

Slobodin, Richard, 1978. *W. H. R. Rivers*. New York: Columbia University Press.

Smiles, Samuel, 1859. *Self Help*. London: John Murray.

Smith, Grafton Elliot, 1924. *The Evolution of Man*. London: Humphrey Milford and Oxford: Oxford University Press.

Smith, John Maynard, 1982. *Evolution Now: A Century after Darwin*. London: Macmillan.

Smith, R., 1972. "Alfred Russel Wallace: Philosophy of Nature and Man." *Brit J. Hist. Sci.* 6:177–99.

Sollas, W. J., 1911. *Ancient Hunters and Their Modern Representatives*. London: Macmillan.

Spencer, Herbert, 1851. *Social Statics; or, the Conditions Essential to Human Happiness Specified*. London: John Chapman.

————, 1855. *Principles of Psychology*. London: Longman.

————, 1862. *First Principles of a New Philosophy*. London: Williams and Norgate.

————, 1864. *Principles of Biology*. 2 vols. London: Williams and Norgate.

————, 1870–71. *Principles of Psychology*. 2d ed. 2 vols. London: Williams and Norgate.

————, 1876–96. *The Principles of Sociology*. 3 vols. London: Williams and Norgate.

————, 1883. *Essays, Scientific, Political, and Speculative*. 3 vols. London: Williams and Norgate.

————, [1884] 1969. *The Man versus the State*. Reprint. Harmondsworth, Eng.: Penguin Books.

————, 1887. *The Factors of Organic Evolution*. London: Williams and Norgate.

————, 1893. "The Inadequacy of Natural Selection." *Contemporary Review* 43:153–66, 439–56.

Stepan, Nancy, 1982. *The Idea of Race in Science: Great Britain, 1800–1960*. London: Macmillan.

Stephens, Lester D., 1982. *Joseph LeConte: Gentle Prophet of Evolution*. Baton Rouge: Louisiana State University Press.

Stocking, George W., Jr., 1962. "Lamarckianism in American Social Science." *J. Hist. Ideas* 23:239–56.

————, 1968. *Race, Culture, and Evolution*. New York: Free Press.

————, ed., 1984. *Functionalism Historicized: Essays on British Social Anthropology*. Madison: University of Wisconsin Press.

————, 1987. *Victorian Anthropology*. New York: Free Press.

Stone, Irving, 1981. *The Origin*. London: Cassell.

Stubbe, Hans, 1972. *History of Genetics*. Cambridge, Mass.: MIT Press.

Sturtevant, A. H., 1965. *A History of Genetics*. New York: Harper and Row.

Sulloway, Frank, 1979a. "Geographical Isolation in Darwin's Thinking: The Vicissitudes of a Crucial Idea." *Stud. Hist. Biol.* 3:23–65.

————, 1979b. *Freud, Biologist of the Mind: Beyond the Psychoanalytic Legend*. London: Burnett Books.

————, 1982. "Darwin and his Finches: The Evolution of a Legend." *J. Hist. Biol.* 15:1–54.

Swinburne, R. G., 1965. "Galton's Law — Formulation and Development." *Annals of Sci.* 21:15–31.

Symondson, A., 1970. *The Victorian Crisis of Faith*. London: SPCK.

Taylor, Gordon Rattray, 1983. *The Great Evolution Mystery*. London: Secker and Warburg.

Temkin, Owsei, 1959. "The Idea of Descent in Post-Romantic German Biology." In *Forerunners of Darwin*, ed. B. Glass et al., 323–55. Baltimore: Johns Hopkins Press.

Tennenbaum, J., 1956. *Race and Reich*. Boston: Twayne.

Tennyson, Alfred, 1894. *The Works of Alfred, Lord Tennyson*. London: Macmillan.

Turner, Frank M., 1974. *Between Science and Religion: Reaction to Scientific Naturalism in Late Victorian England.* New Haven: Yale University Press.

———, 1978. "The Victorian Conflict between Science and Religion: A Professional Dimension." *Isis* 69:356–76.

Turrill, W. B., 1963. *Joseph Dalton Hooker.* London: Scientific Book Guild.

Tylor, Edward B. [1865] 1870. *Researches into the Early History of Mankind.* 2d ed. London: John Murray.

———, 1881. *Anthropology.* London: Macmillan.

Vorzimmer, Peter J., 1970. *Charles Darwin: The Years of Controversy.* Philadelphia: Temple University Press.

Wagner, Mortiz, 1873. *The Darwinian Theory and the Law of the Migration of Organisms.* London: E. Stanford.

Wall, Joseph, 1970. *Andrew Carnegie.* New York: Oxford University Press.

Wallace, Alfred Russell, 1855. "On the Law Which has Regulated the Introduction of New Species." *Ann. & Mag. of Nat. Hist.* 26:184–96.

———, [1869] 1962. *The Malay Archipelago.* Reprint. New York: Dover.

———, 1870. *Contributions to the Theory of Natural Selection.* London: Macmillan.

———, 1876. *The Geographical Distribution of Animals.* 2 vols. London: Macmillan.

———, 1880. *Island Life.* London: Macmillan.

———, 1889. *Darwinism: An Exposition of the Theory of Natural Selection.* London: Macmillan.

Ward, Lester Frank, 1895. "The Relation of Sociology to Anthropology." *Am. Anthropologist* 8:241–56.

Wassersug, R. J., and M. R. Rose, 1984. "A Reader's Guide and Retrospective to the 1982 Darwin Centennial." *Quart. Rev. Biol.* 59:417–47.

Weiner, J. S., 1955. *The Piltdown Forgery.* London: Oxford University Press.

Weismann, August, 1880–82. *Studies in the Theory of Descent.* 2 vols. London: Sampson Low.

———, 1891–92. *Essays upon Heredity and Kindred Biological Problems.* 2 vols. Oxford: Oxford University Press.

———, 1893a. *The Germ Plasm: A Theory of Heredity.* London: Scott.

———, 1893b. "The All-Sufficiency of Natural Selection." *Contemporary Rev.* 64:309–38, 596–610.

———, 1896. *On Germinal Selection.* Chicago: Open Court.

———, 1904. *The Evolution Theory.* 2 vols. London: Arnold.

Weiss, Sheila F., 1986. "Wilhelm Schallmayer and the Logic of German Eugenics." *Isis* 77:33–46.

Weldon, W. F. R., 1894–95. "An Attempt to Measure the Death-Rate due to the Selective Destruction of Carcinas moenas." *Proc. Roy. Soc. Lond.* 57:360–79.

———, 1898. "President's Address, Zoology Section." *Report of The Brit-*

ish Association for the Advancement of Science, 1898 meeting, 887–902.

———, 1901. "A First Study of Natural Selection in Clausilia laminata." Biometrika 1:109–24.

Wells, Kentwood D., 1973a. "William Charles Wells and the Races of Man." Isis 64:215–25.

———, 1973b. "The Historical Context of Natural Selection: The Case of Patrick Matthew." J. Hist. Biol. 6:225–58.

White, A. D. [1896] 1960. A History of the Warfare of Science with Theology in Christendom. 2 vols. Reprint. New York: Dover.

Whitehead, A. N., 1929. Process and Reality: An Essay in Cosmology. Cambridge: Cambridge University Press.

Willey, Basil, 1956. More Nineteenth-Century Studies. London: Chatto and Windus.

———, 1960. Darwin and Butler: Two Versions of Evolution. London: Chatto and Windus.

Williams-Ellis, Amabel, 1969. Darwin's Moon: A Biography of Alfred Russel Wallace. London: Blackie.

Wilson, Edward O., 1975. Sociobiology: The New Synthesis. Cambridge, Mass.: Harvard University Press.

———, 1978. On Human Nature. Cambridge, Mass.: Harvard University Press.

Wilson, Raymond J., 1967. Darwinism and the American Intellectual: A Book of Readings. Homewood, Ill.: Dorsey Press.

Winsor, Mary P., 1979. "Louis Agassiz and the Species Question." Stud. Hist. Biol. 3:89–117.

Wood, Robert Muir, 1985. The Dark Side of the Earth. London: Allen and Unwin.

Woodcock, George, 1969. Henry Walter Bates: Naturalist of the Amazons. London: Faber and Faber.

Woodward, Arthur Smith, 1915. A Guide to the Fossil Remains of Man. London: British Museum (Natural History).

Wright, Sewall, 1930. "The Genetical Theory of Natural Selection." J. Heredity 21:349–56.

———, 1931. "Evolution in Mendelian Populations." Genetics 16:97–159.

Wyllie, Irvin G., 1954. The Self-Made Man in America. New Brunswick, N.J.: Rutgers University Press.

———, 1959. "Social Darwinism and the Businessman." Proc. Am. Phil. Soc. 103:629–35.

Yeats, W. B. [1933] 1950. The Collected Poems of William Butler Yeats. 2d ed. London: Macmillan.

Young, Robert M., 1969. "Malthus and the Evolutionists: The Common Context of Biological and Social Theory." Past and Present 43:109–45.

———, 1970. Mind, Brain, and Adaptation in the Nineteenth Century. Oxford: Clarendon Press.

————, 1971. "Darwin's Metaphor: Does Nature Select?" *The Monist* 55:442–503.

————, 1973. "The Historiographical and Ideological Context of the Nineteenth-Century Debate on Man's Place in Nature." In *Changing Perspectives in the History of Science*, ed. M. Teich and R. M. Young, 344–428. London: Heinemann.

————, 1985a. *Darwin's Metaphor: Nature's Place in Victorian Culture.* Cambridge: Cambridge University Press.

————, 1985b. "Darwinism *Is* Social." In *The Darwinian Heritage*, ed. David Kohn, 609–38. Princeton: Princeton University Press.

Zmarzlik, Gunter, 1972. "Social Darwinism in Germany." In *Republic to Reich: The Making of the Nazi Revolution*, ed. Hajo Holborn, 435–74. New York: Pantheon.

INDEX

The Non-Darwinian Revolution

Designed by Ann Walston.

Composed by Capitol Communication Systems, Inc.
in Trump text and display type.

Printed by the Maple Press Company
on 50-lb. S. D. Warren's Sebago Eggshell Cream offset
and bound in Joanna Arrestox
with Rainbow Antique endsheets.

Printed in the United States
208637BV00002B/268-273/A

9 780801 843679